Writing Successful Science Proposals

Writing Successful Science Proposals

SECOND EDITION

Andrew J. Friedland and Carol L. Folt

YALE UNIVERSITY PRESS New Haven & London

First edition published 2000. Second edition published 2009.

Designed by James J. Johnson and set in Scala Roman type by The Composing Room of Michigan, Inc.
Printed in the United States of America.

Library of Congress Cataloging-in-Publication Data

Friedland, Andrew J., 1959–
    Writing successful science proposals / Andrew J. Friedland and Carol L. Folt. — 2nd ed.
        p. cm.
    Includes bibliographical references and index.
    ISBN 978-0-300-11939-8 (pbk. : alk. paper)
    1. Proposal writing in research.   I. Folt, Carol L., 1951–
II. Title.
    Q180.55.P7F75 2009
    001.4′4—dc22

                                                2008046322

A catalogue record for this book is available from the British Library.

This paper meets the requirements of ANSI/NISO Z39.48–1992 (Permanence of Paper). It contains 30 percent postconsumer waste (PCW) and is certified by the Forest Stewardship Council (FSC).

10   9   8   7   6   5   4   3   2   1

Contents

Preface to the Second Edition

After more than a decade of teaching a course on proposal development and writing, we believe that the need for reference works on proposal writing for students and researchers has grown even stronger. Not only is the ability to write successful research proposals important for the modern scientist, but introducing students to the process of "doing science" as early as possible in their training improves the quality of their learning. The NIH, NSF, and other agencies have actively encouraged the insertion of scientific inquiry and research development into the contemporary undergraduate science curriculum. Although there is no single formula for a strong proposal, there are a number of precepts that are consistently found in successful proposals, and the sooner one learns these precepts, the better.

We've been very thankful for the feedback we've

received on the first edition of this book. Readers will notice important changes and updates on electronic submissions, and new and revised text describing current federal criteria and formatting. We've also included a new chapter on writing proposals for private foundations, and another new chapter on multidisciplinary, multi-investigator proposals. In a number of locations, we've suggested ways that investigators can increase the impact of their research by seeking matching funds, and can increase the value of their proposals and funded work. We hope you will agree that this second edition adds valuable new information and examples while maintaining the brevity of the first edition.

Preface

One of the most challenging aspects of scientific research is synthesizing past work, current findings, and new hypotheses into research proposals for future investigations. Such research proposals combine every aspect of scientific inquiry, from the creative conceptualization to the detailed design, projected analysis of the data, synthesis of the results, and estimation of the budget. Because grant applications are an articulation of the scientific process, writing them is one of the most exciting parts of "doing science." If you are planning to write a grant application for a major foundation, such as the National Science Foundation, the Environmental Protection Agency, or perhaps a private foundation, or if you are writing a proposal to conduct research as a graduate student or undergraduate, this book should be of value to you.

Many research institutions offer graduate-level

courses on proposal development, and research design
is growing increasingly vital in the undergraduate science curriculum. Given the importance of this subject
to future scientists, our faculty in ecology and environmental studies at Dartmouth College felt that it was essential that we create a course on scientific project design and proposal writing for our graduate students. In
1994, when we began teaching the course, we could
not find a text that specifically addressed grant writing
in the natural sciences. So we decided to write one our-
selves based on our experiences in the classroom. We
hope that our book will be of value not only to students
but also to new researchers seeking to improve their
skills in developing research proposals.

This book provides guidance for those conceptualizing and formulating their research plans, and it
offers specific instruction on organizing and presenting material in a standard format. We offer an overall
organizational framework, and we list the components
of successful scientific proposals. Before you begin to
write, you must have a very clear idea or concept for
your research. There is, however, no secret formula for
writing such proposals. Each grant application must be
tailored to the specifications of the funding agency or
graduate committee to which it is directed.

Research proposals are written for a variety of

purposes and are submitted to many different agencies and to faculty committees. We focus on agencies that solicit proposals in the natural sciences; these include the National Science Foundation (NSF), National Institutes of Health (NIH), Environmental Protection Agency (EPA), U.S. Forest Service (USFS), U.S. Geological Survey (USGS), and private corporations and foundations, as well as academic committees. Our format should also be useful to those submitting to the National Research Council of Canada, NATO Scientific and Environmental Affairs Division, and other funding agencies worldwide.

There are many ways to write excellent proposals. We present a model that we and our students and colleagues have used with success. Our ideas have been combined with those of the many natural scientists from a variety of disciplines with whom we have consulted while writing this book. Discussions with colleagues, proposals given to us by successful authors in a variety of fields, and our students' ideas have been especially meaningful in this effort. If you submit a proposal after using this book, or if you use this book in a course, please let us know how you fare. We look forward to hearing from you.

Acknowledgments

We are grateful to the many students, colleagues, advisers, reviewers, and program managers who have contributed greatly to our proposals over the years or directly to this project. While we were writing this book, numerous people generously shared ideas, experiences, and proposals with us. We hope that the following list includes most of the people with whom we have communicated. Our sincere apologies for any omissions: John Aber, Victor Ambros, Matt Ayres, Joel Blum, Doug Bolger, Christine Bothe, Rick Boyce, C. Page Chamberlain, Celia Chen, Ann Clark, Jim Coleman, Hany Farid, Marcelo Gleiser, Mary Lou Guerinot, Nelson Hairston, Jr., Dick Holmes, Mary Hudson, Tom Jack, Kevin Kirk, Jon Kull, Eric Lambie, Jane Lipson, Pat McDowell, Mark McPeek, Frank Magilligan, Eric Miller, William North, Jerry Nunnally, George O'Toole, Bonnie Paton, David Peart, Bill Reiners, Jim Reynolds,

Roger Smith, Richard Stemberger, Judy Stern, Amy
Stockman, Ross Virginia, Richard Wright, Wayne
Wurtsbaugh, Ruth Yanai, and four anonymous review-
ers.

Our special recognition goes to our dear colleagues, Noel Perrin and Donella Meadows, both of
whom have passed away since the first edition was
published. Their advice for navigating the publishing
world was invaluable. Lisa Clay, Carrie Larabee, and especially Susan Milord contributed a range of ideas, as
well as editing and technical assistance. Margaret Dyer
Chamberlain provided many cartoons for our consideration. Finally, we thank David Peart, Noel Perrin, and
two anonymous reviewers for carefully reading versions of the original manuscript, Heidi Downey for
valuable editorial assistance, Matthew Laird for logisti-
cal assistance, and Jean Thomson Black for her support, enthusiasm, and hard work as our editor.

We also wish to acknowledge the many users of
our book who have contributed important input and
encouragement. One of the most resounding comments was to keep the second edition brief! We hope
we have managed to follow this advice.

A Note to the Reader

We wrote this book to be read two ways: either in its entirety before beginning a project or chapter by chapter—but not necessarily in sequence—as you develop specific sections of your proposal. The following list contains a number of goals that you can realistically expect to accomplish over the course of preparing a research proposal.

- Identify and describe the conceptual framework for the research question.

- Review the relevant theoretical and empirical literature both for the system being studied and for related systems.

- Describe the general research question in the context of the conceptual framework and the theoretical and empirical work that precedes the proposed work.

- Formulate a concise and incisive set of hypotheses or specific aims to address the overarching question.

- Design studies to test each hypothesis or aim.

- Develop methods and techniques to test, analyze, and synthesize results.

- Evaluate potential alternative outcomes that may be obtained from each part of a study, and consider where each of these alternatives may lead.

- Combine these items in a coherent, precise, concise, exciting proposal.

- Submit the proposal to the appropriate agency or evaluation committee.

- Interpret and respond constructively to reviews of the proposal.

This primer contains a collection of chapters that address our dual goals of assisting with the development of research ideas and of providing detailed guidelines for writing grant applications. We present the material in much the same order we use in teaching our course, Design and Development of Scientific Proposals, and in designing our own research propos-

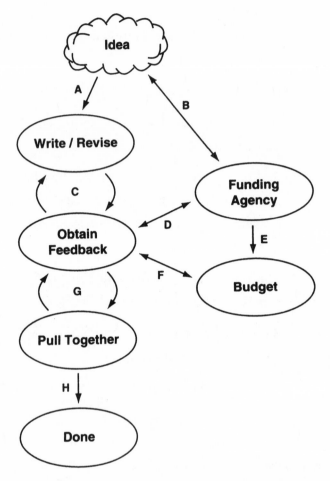

An idea begins the proposal development and writing process, but sometimes a request for proposals from a particular agency can influence or motivate a project. From the original idea in its written form, scientists, funding agency personnel, and agency or budgetary guidelines and restrictions can provide inputs that the writer uses to revise the proposal. After reconciling all comments and feedback, the author submits the final document.

als. We first discuss general types of proposals and
share thoughts about writing research grants applications (Chapters 1 and 2), and then we outline the basic
elements of a proposal and the various proposal types
(Chapter 3). We consider the conceptual framework
(Chapter 4) and how and where in the proposal to articulate succinctly the study's importance. In Chapters 5–
13, we address the requirements and construction of
the specific elements found in most standard proposals (summary, background, methods, budget). We present the mechanics of submitting and tracking a proposal and revising and resubmitting in Chapters 14
and 15. In Chapters 16 and 17 (new to the second edition), we present a discussion of submitting proposals
to private foundations, and we discuss the particular
needs of multidisciplinary, multi-investigator proposals. We conclude by sharing our thoughts about ethics
and scientific research (Chapter 18).

CHAPTER 1

Getting Started

The need to develop precise, concise, well-articulated grant applications has become more important than ever in the past few years. The increase in specific requirements for proposals and the greater scrutiny they receive have made proposal writing more demanding; we hope it will continue to be an enjoyable and educational experience for new as well as experienced proposal writers.

For some scientists, designing research carries the same sense of exploration, anticipation, and unlimited opportunity as the first day of a new school year. For this reason, it can become a scientist's favorite endeavor. As you begin your research proposal, we urge you to:

THINK BIG. Reflect on your question from its broadest perspective. Imagine finding innovative solutions to fundamentally important problems. If you start small, your work will end up even smaller.

AVOID TUNNEL VISION. Consider projects that could lead to years of research. Enjoy a time of intense creativity, and—at least for a while—think beyond your immediate research area.

DREAM. Dream about solving important problems, making a difference, producing significant papers, and making discoveries.

TAKE YOUR TIME. Great ideas do not appear in thirty-minute windows of time. When designing a research project, expect to spend lots of time on it. You will.

Planning research can be stressful. Anxiety arises when we focus too much on what people will think of our work. We all have periods of insecurity, times when we mistakenly believe that everything rests on the outcome of one specific project. People often fret about how their advisers or peers will evaluate them. They worry about their research questions: "Will I think of a question important enough to keep my interest and warrant my attention for years to come?" They feel uncertain about the outcome: "Will my research idea work?" "Will it lead to publications?" Try not to be overly concerned. Many people experience anxiety when they feel pressured to identify research problems.

Reducing the uncertainty associated with developing a scientific proposal fosters the excitement and innovation that lie at the heart of science and research design. Here are some simple steps to ease yourself into the process:

- *Define tasks associated with the proposal.* Don't make the list too long or too wide-ranging at the start, or it will be discouraging.

- *Develop a timeline or strategy for working on your proposal.* Try working backward from your deadline to get a reasonable idea about when specific tasks must be accomplished. Make sure that you have sufficient time.

- *Accomplish something early.* Complete a few tasks quickly. We give our class a set of short- and long-term deadlines at the start of the term. (We'll present a few examples later in this chapter.)

- *Remember that the best proposals are built from the best science. Effective proposals require a sound scientific basis.* Developing and then articulating a logical framework for the problem are the key elements in the success and power of a research proposal. Therefore, time spent developing ideas is well spent. Some researchers believe that the best problem solvers are individuals who understand the need to get the initial question right (which is often difficult to do) (Runco 1994, Proctor 2005).

- *Relax, and be prepared for change.* Nothing is fixed. You will think and rethink everything as you develop your proposal.

Exercises for Getting Started

We use three exercises to initiate proposal development. These tasks are not meant to be accomplished in a single sitting but should be pursued concurrently: critique other proposals; accomplish administrative and technical tasks; work on the conceptual framework of your research.

Critique other proposals. Established scientists routinely review the proposals of students and colleagues as part of the peer review process. This gives them a sense of the scope and size of the best research proposals. Assessing other research proposals is also a

potent method of learning science and focusing on
both the broad implications and the methodology behind research. It is general policy that reviewers destroy proposals after reading them, but most scientists
will share their own successful and unsuccessful proposals with peers and students. Do not hesitate to request such assistance from colleagues.

As you read proposals, consider the following
major criteria: scientific content, innovation and scope
of ideas and methods, structure and format, clarity, and
style. Reviewers for the National Science Foundation or
other granting agencies may be asked to consider the
following while evaluating a proposal: the scientific importance of the question(s), the broader implications of
the research to the community at large, the rigor of hypotheses or testability of the study questions, the feasibility of research design, the qualifications of the investigator, and the suitability of facilities for the proposed
work.

Our class begins with a discussion of proposals
that we have written or that have been given to us by
our colleagues to share with the group. Using the title,
project summary (or abstract), and significance sections, we ask the class to question whether the author
has convincingly justified the proposed work. We discuss methods, graphics, and style and ask whether the

work captured our attention. At some point we try to compare each proposal with others we have read. This discussion is meant to be a starting point; eventually everyone develops individual styles, methods, and measures for evaluating proposals.

Accomplish administrative tasks. Completing administrative and technical tasks is another effective way to get started. Begin by reading the proposal guidelines and requirements for the potential funding agency or foundation, or the guidelines issued by your department. Fairly early in the process you should put together a simple outline identifying the key sections of the final document (see Chapter 3). Think about the optimal length for each section. This activity will probably put you at ease because you will quickly realize that most grant applications are usually concise—fifteen single-spaced pages is the maximum for NSF; many other agencies have the same page limit, whereas others, such as dissertation improvement grants offered by some NSF programs, are much shorter.

Another important task is to determine an institution's procedures for grant processing. Ask questions such as, "What paperwork must I complete?" "What signatures do I need?" "Where do I go?" "How much time should I allow?" "What are the institution's

rules on budgets, overhead costs, and cost sharing?"
"Do I need special permission for anything?" (e.g., animal care, use of human subjects). These seemingly
mundane points are critical, as poor planning may result in a scramble to meet due dates, or, worse, missed
deadlines. Speak to your institutional grant manager as
early as possible. Getting advice from an experienced
colleague will also save you much time in the long run.

Most granting agencies have already or are in
the process of moving to electronic submission and
processing of grant applications. For example, most
U.S. NSF programs require electronic submission
through a program called FastLane (www.fastlane.nsf
.gov). Some NSF programs and most other federal
grant programs receive grant applications through
Grants.gov (www.grants.gov), which claims to be the
primary access point for electronic applications to
more than one hundred programs from all twenty-six
federal granting agencies. These programs allow a researcher to use the Web to prepare and submit the proposal—including figures, tables, and budget—and
they allow for multiple authors to work on the same
document. Electronic submission saves on expensive
processing and paperwork and negates the need for
physical delivery of the proposal. As you collect information for submitting a proposal to your potential

funding agency, be sure to learn about the requirements regarding electronic submission.

    Develop the conceptual framework for your project.
Conceptualizing your research is the most substantial
step in preparing a proposal. Some people work on
their ideas for months or years before they actually begin to write. Others, especially students, pull together
their ideas only when they are required to write their
first research proposal. In our proposal development
class we spend several weeks working to produce a succinct statement of the overall concept that can be understood by a broad scientific audience (see Chapters
4–7). This statement is the foundation for the rest of
the proposal (see Chapters 8 and 9).

Know Your Audience

    Grant applications are written for a variety of
purposes and are submitted to many different types of
agencies. Before you begin writing, consider the fit between your research goals and the targeted agency.
Agencies have various reasons for announcing a Request for Proposals (RFP) or establishing a program
that will periodically accept proposals. In this book we
focus on federal funding agencies, such as the NSF

LITZLER

"WAS THERE ANY TALK ABOUT FUNDING?"

and the NIH, and private corporations and foundations. For the most part, we discuss basic research proposals in which the investigator sets out research questions and goals. Agencies sometimes set the goals, however, and request proposals to address a particular objective, research target, or initiative. Accordingly, we separate proposals into two general categories:

1. Basic research proposals (unsolicited research proposals), which generally must provide novel insights or methodologies for solving fundamental scientific problems (see Chapter 4).

2. Task-oriented or program-initiated proposals, in which the topic or goal of research is specified by an agency, a corporation, or a foundation.

There is usually less latitude in determining research topics than for basic research proposals. Proposals are evaluated on their likelihood of accomplishing the specified task, so emphasis is placed on methods, ability to accomplish the project, credentials, the projected outputs, and time needed to complete the project. (These criteria are also important in basic research proposals.)

Both categories of proposals are usually solicited by a long-standing or newly announced Request for Proposals (RFP) or Request for Applications (RFA) or another announcement of intent. RFPs have specific language that indicates what types of activities will be funded by the specific agency. It is essential to read the RFP carefully and talk to colleagues who have experience with that particular agency before proceeding too far in your planning.

Once you have identified a specific program or agency and become familiar with the RFP and other guidelines, talk with the program director or manager (the person in charge of evaluating grants in that program). Do not call until you have definite questions. Avoid open-ended queries, such as, "What kind of proposals do you fund?" Take notes during the conversation. Discuss the goals and general format of your project and ask such questions as, "Does my proposed research fit within the mandate of your program?" "Is there a related program that you think would be better suited to evaluate my project?" There also may be unwritten requirements for successful grants that you need to clarify with the program director. For example, you may wish to address a question by comparing data from diverse regions around the world, but the agency may be interested only in questions about a particular region. The program manager can explain such issues relating to the scope of the program. Be sure to ask about spending limits, restrictions on equipment purchases and investigator salaries, and other financial regulations. (See also Chapter 13.)

It is also appropriate to ask the program manager about the review process. Find out the general types of backgrounds of the scientists who will evaluate your application. By knowing your audience, you can anticipate questions and address likely concerns in the proposal. For proposals that cross disciplinary boundaries, this information is especially critical. When conducting interdisciplinary research you will need to address the questions of individuals in each discipline. Discussions with the program director and with scientists in the pertinent fields will save you much time and effort and could make the difference between success and failure.

Other Exercises for Getting Started

- Distinguish tasks that can be accomplished in one or two days from longer-term chores.

- Find at least one set of proposal guidelines. This can be accomplished by contacting the

LITZLER

"AS WE CONSIDER THE PROPOSALS, LET'S NOT FORGET WHO INCLUDED A GALLON OF ROCKY ROAD AS 'ATTACHMENT B'."

office on your campus that handles the submission and administration of grant awards, surfing the Web for agencies' guidelines, or borrowing from a colleague or an adviser. See the Web addresses for funding agencies in appendix 2.

• Begin to identify specific sections required in the final proposal and to list the elements they should contain.

• Download abstracts of funded proposals from agency Websites.

C H A P T E R   2

Authorship from Start to Finish

Responsibility for research extends from conception and completion of a proposal to publication and future use of the resulting data. The individuals who accept this responsibility—and the credit for the ideas, methodologies, and eventual results developed in a proposal—are the authors of that proposal. Sometimes authorship is shared, and coauthorship of grants usually leads to coauthorship of the resulting publications. Because students often find authorship a difficult and uncertain topic, we discuss it near the start of our proposal development class.

There are two essential points to consider about authorship:

- *Appreciate all that goes into research.* Understand the components of designing, conducting, analyzing, and writing scientific research. Fully acknowledge sources of supporting information, derivative ideas, and collaborators. As this understanding and appreciation develop, authorship decisions will be made appropriately, and there will be fewer chances for misunderstanding.

• *Discuss expectations.* Determine in advance specific expectations for each author and collaborator for all phases of proposal development and research implementation. This includes authorship and revision of publications resulting from a research project.

Define Expectations at the Outset

Scientific advances are best achieved through honest sharing of ideas, constructive critical appraisal by colleagues, and revisions in approaches and perspectives resulting from these debates. Science is always collaborative at some level, and as such it requires trust and understanding among associates. Be generous with your feedback to others and you will benefit as well. Don't let fears of having your ideas taken without credit diminish your relationships with collaborators and colleagues. The most effective strategy for avoiding problems over collaboration and authorship is to establish, at the start, clear expectations for all members of a project.

Who is the first author and what does the first author do? Every author on a proposal or paper has assumed intellectual, ethical, and fiscal responsibility for the research project. The first author on a funded project, the principal investigator (PI), takes primary responsibility for the project, just as the first author of a scientific paper usually takes primary responsibility for the overall design and execution of work and interpretation of results (cf. Day 2006). This author often does the greatest amount of work conceptualizing and implementing the project, although some projects may be conceived by one person and carried out and expanded by another. As Day points out, accepted conventions for the order of authors (and, by extension, investigators) vary by field. The PI on a proposal might become the last author on a publication, depending on the conventions of the particular field. Nowadays, many journals request information about each author and what he or she contributed (conceived the question, conducted experiments, analyzed data, wrote the manuscript).

For certain proposals, the first author and presence or absence of collaborators are mandated by the funding agency or review committee. For example, proposals submitted by graduate students to dissertation committees usually carry only the student's name even though the adviser usually has a guiding role in the proposal. A later application to an outside agency may well be cowritten by the student and the adviser. Some funding programs (such as NSF dissertation improvement grants) require that the PI on grants be a faculty member at a research institution or an accredited college or university. This policy ensures that applicants have scientific training and experience and affiliation with an institution that is committed to management of the work; it also directs fiscal and ethical responsibility to a defined individual and place. These types of requirements are usually discussed under a section describing eligibility for specific grants.

Sources versus collaborators. Aside from "obvious" associations (e.g., between advisers and students, between long-standing research collaborators, and between co-writers), co-investigators are usually added to a proposal when their expertise is required for the research. If you have a casual conversation with someone and provide a good idea that he or she later uses, you should be flattered. You may be acknowledged, but your suggestion will probably not entitle you to participate in the project. Interactive scientists make suggestions about other people's research all the time; if your suggestions are good, they will be used frequently, but they may seldom lead to collaboration.

Being a co-investigator on a grant application generally implies a long-term contribution to the ideas, design, implementation, data analysis, and future publications. Each case should be decided on its own merits, but early discussion of this issue is essential.

Appreciate All That Goes into Research, and Give Credit Where Credit Is Due

You should discuss authorship of possible publications even as the proposal is being written. Inherent in such discussion are expectations that the research will be taken seriously, be executed well, and be interesting enough to merit publication. You and your collaborators will probably return to this discussion several times after the research design is complete.

To give proper credit and to determine authorship of eventual publications, you must recognize the importance of all contributions to the scientific research. This understanding is also critical for writing a proposal and determining how and when tasks will be performed. Less-experienced researchers may feel that the most important contributions come from those who physically collect the data. But even though data collection is important, participation in the following areas may be even more critical to the outcome of the research, and it could lead to first authorship of eventual papers:

- identification of research topics and important problems and questions

- formulation of the theoretical foundations of the research

- design of effective research protocols

- sample analysis

- data analysis

- writing of the document

If you and your collaborators discuss authorship early you will avoid misunderstandings later. This discussion may also help guide decisions about who will do key pieces of the project and about the timeline for project completion, which will facilitate the success of the research effort.

No One Owns Ideas. Right?

Research is not about ownership, but about sharing. Yet we all have assumptions regarding who gets to use project data and publish the results. Normally the investigators of a project have sole access to the first use and publication of data. After results are published, they may be used by anyone who properly cites the published work. There has been an increasing trend for funding agencies to require that investigators make data publicly accessible via the Web or upon request. Some agencies require that all data be made available within a certain time period, even if the data have not been published.

Different disciplines have different protocols and practices. Laws regarding intellectual property rights and patents apply in some areas of science and not in others. If you have questions about the ownership of ideas, techniques, or instruments you plan to develop, be sure to contact your institution's grants administration office before you move too far into the project. Difficulties can arise if collaborators have not anticipated what will happen to research and data after a proposed project is completed. It is particularly important to discuss this issue with students, postdoc-

toral fellows, and other research associates before beginning a collaborative project. The following examples illustrate common situations in which research
may be expected to "stay within a lab" (i.e., under a specific laboratory director or PI) after the primary researcher (e.g., a graduate student, technician, research
associate) on that project leaves the lab.

    Longitudinal studies (studies that follow specific
individuals or sites over time). The value of longitudi-
nal studies grows over time, and the initial data base often yields additional research. Examples of longitudi-
nal data bases include observation of individuals after
exposure to a variable (e.g., pollutants, drugs, hazards);
observation of plots of trees as they grow from seedling
stage through maturity; repeated analysis of cancer
registries; or repeated visits (perhaps once per decade)
to sample soils or sediments. In most cases, the investigators who designed, arranged to fund, and executed
an original study control access to it after the initial
grant period by continuing to maintain the research
sites or follow the individuals involved in the study. A
student or co-investigator who joins a project with specific objectives for use of the original sites or individuals should not assume that future access is guaranteed.

    Laboratory study systems. In some fields it is
common for students and postdoctoral fellows to con-

duct research on systems (e.g., specific genes or gene products) that have been identified, described, and manufactured by previous "generations" of scientists in the same lab. Scientists training in these laboratories often leave their project behind when they depart from a lab; they develop new systems (separate from their thesis work) when they assume positions elsewhere.

Technique and instrumentation development. For many scientists, developing new techniques and applying sophisticated instrumentation to novel problems are primary research objectives. Individuals who use these techniques or rely on instrumentation developed and maintained by others cannot assume that they will have future access to and involvement with the instrument or techniques.

We urge you to discuss this issue in advance with advisers, colleagues, and collaborators. Understand the culture of each field with respect to this concept. If you are uncomfortable with the prospects, discuss them with the appropriate people early in the project. Don't assume that you can agree now and change the expectations later.

One of us was involved with five graduate students and faculty members in a large project. Early in the project we wrote down the area in which each person would assume responsibility for doing research, writing, and revising. We even wrote the potential titles of resulting papers and listed the journals in which we intended to publish. Five years later, much of the work had been completed and most of the papers had been written, and there was little debate about who would be the lead author on each paper. We ascribed this consensus, in part, to our identifying the "ownership" of the different parts of the project early in its development.

Exercises for Authorship

The best preparation for thinking about authorship of proposals and papers is to share opinions about this topic with your adviser or colleagues. We have an annual seminar with our graduate students on this issue, and it is always thought-provoking and interesting for the participants.

Read the following sequence of scenarios and discuss your expectations for collaboration or eventual authorship.

> • You have developed a set of hypotheses to test in a system you know quite well. You present your ideas at an informal department gather-

ing. A colleague brings to your attention a particular paper that describes an appropriate method. What should you do? If you use the method in your eventual design, what does your colleague expect in the way of recognition?

• Now, consider a situation in which the methodology needed for conducting your tests does not exist or requires a method being developed by a colleague. In this situation, suppose the colleague suggests a way to achieve your goal, or states that she could develop a method that would meet your needs. What should you do? If, in your eventual design, you use the method she develops, what do you and your colleague expect about her recognition?

• Finally, consider a situation in which you develop a series of ideas on how to test your hypotheses as a result of a number of late-night and hallway chats held over weeks or months with a particular colleague. What if your colleague's perception of how much help you received differs from yours? In this scenario, a misunderstanding may easily occur. What should you do?

Basic Organization and Effective Communication

Many novice writers find it difficult to decide on an organizational structure for their proposal. The number of sections and the disparate types of information that must be included can be overwhelming. Although the key to a good proposal is sound science, efficient organization makes a scientifically convincing project even stronger.

Some funding agencies are flexible in their specifications for proposal format. Others require that sections be presented in a particular sequence. Our recommendations in this chapter are based on the format suggested by the U.S. National Science Foundation in the Grant Proposal Guide (currently referred to by NSF as GPG, NSF 08–1 [January 2008]; be sure to use the most up-to-date version).

Five Precepts for Effective Organization and Communication

Effective communication substantially improves your chances of success. If your language is clear and precise, and your document well organized, your ideas will be better understood, their importance will be more apparent, and the comments of reviewers will be more useful. We propose five axioms for communicating your ideas:

- Organize
- Highlight
- Funnel
- Focus
- Unify

A well-organized document is easier to follow and comprehend. Highlight your most important points early in your proposal. This directs the reader toward the issues that you feel are vital and thereby increases your impact. Do not emphasize ideas that are less important to your research, and do not bury critical information. Whenever possible, funnel the reader from the big picture to the specifics of your research. Focus on that topic—avoid information that detracts from or dilutes your message. Funneling and focusing establish for the reviewer that your project is the most effective and logical way to answer the questions you have raised. Offer the readers a "road map" early in the proposal (and often in each section) to keep them headed in the direction you wish. Finally, unify the voice and central goals of your project (see Chapter 17 for further discussion of the importance of unifying).

Organizing a Proposal for the NSF

The National Science Foundation is an independent federal agency mandated to promote and advance scientific and engineering progress in the United States. Earlier this decade, the foundation received more than forty thousand proposals annually and authorized about eleven thousand new awards each year (NSF Grant Proposal Guide, 2004, p. 2). However, many programs within NSF fund 20 percent or fewer of the proposals they receive, so the competition is intense.

The GPG, which is free, is a very useful document for proposal writing. The guide is valuable for anyone working on a proposal, even if a proposal is not for submission to NSF, because it clearly specifies typical research proposal expectations. For a copy of the GPG, access NSF on the Web at http://www.nsf.gov.

Many types of proposals are submitted to NSF, but we focus here on the so-called Full Proposals. These are basic research proposals that can be submitted to NSF in topic areas, called directorates, from geosciences, mathematical and physical sciences, engineering, biological sciences, and social, behavioral, and economic sciences. Within these and other directorates are perhaps 250 programs (for example, within the directorate for biological sciences there is a division of molecular and cellular biosciences and a program called cell biology; within the directorate for geosciences is a division of earth sciences and a program on hydrologic sciences).

The NSF tells you that the "proposal should present the: (1) objectives and scientific, engineering or educational significance of the proposed work, (2) suitability of the methods to be employed, (3) qualifications of the investigator and the grantee organization, (4) effects of the activity on the infrastructure of science, engineering and education, and (5) amount of funding required. It should present the merits of the proposed project clearly and should be prepared with the care and thoroughness of a paper submitted for publication" (GPG, p. I-4).

Note that the NSF's statement does not identify specific sections or a particular order for the proposal. It does, however, highlight key criteria that NSF reviewers use to evaluate applications:

- General implications of your research (item 1)

- The strength of your argument for funding (items 1 and 5)

- Scientific soundness, fundamental importance, possibility for far-reaching impact, investigator qualifications, and strong likelihood that the project will advance research efforts in a discipline

When submitting their reviews, all reviewers for NSF are asked to justify their evaluations and recommendations with regard to two specific merit criteria (GPG, pg III-1):

Merit criterion 1. What is the intellectual merit of the proposed activity? Reviewers are asked to determine how well the proposed work advances current knowledge in the field, to comment on the qualifications of investigators to conduct the work, and to determine if the proposed ideas are creative, original and worthy of support relative to other proposals also seeking funding.

Merit criterion 2. What are the broader impacts of
the proposed activity? Reviewers are asked to discuss how well the proposed activity advances
"discovery and understanding while promoting
teaching, training, and learning." Reviewers are
also asked how the proposed project broadens
the "participation of underrepresented groups"
and to what extent it will enhance facilities, instrumentation, networks, and partnerships. Reviewers are also asked to comment on the way
results will be disseminated broadly and what
the benefits to society will be.

A "typical" NSF proposal includes the following sections in this general order:

   1. Project summary (or abstract)

   2. Table of contents

   3. Project description (this is the main body of
      the proposal; order is not specified by NSF)

   4. Reference list

   5. Biographical sketches of the investigator(s)

   6. Budget

   7. Current and pending support of the investiga-
      tors

8. Description of the facilities, equipment, and
   other resources available for use

Other organizations may have different specific
requirements, but most include roughly the same elements. The balance of the proposal can vary among
agencies with respect to detail or emphasis on sections
such as experimental protocol. Moreover, a few agencies require additional components, such as quality as-
surance information, special permits, or cooperative
agreements. You will need to obtain this information
from the program director.

Some programs require particular headings
and a fixed order of presentation. For NSF, the overall
sequence is prescribed, but within the body of the proposal the investigator has much latitude concerning
the actual contents, names, and placement of sections.
If your proposal is a resubmission, many review panels
also suggest or require a "resubmission response" in
the main body of the proposal (see Chapter 15). These
sections are usually very helpful.

An outline of the sections commonly found in
proposals is shown below (detailed descriptions are in
upcoming chapters). Some writers do not use all of
these sections, whereas others use more. Ask colleagues or advisers who have received funding to let

you read their successful proposals. Parenthetical references are to chapters in this book.

Project Summary (Chapter 6)
Table of Contents
I. Project Description
   A. Results from prior agency support (Chapter 8)
   B. Statement of the problem and significance (Chapter 4)
   C. Introduction and background (Chapter 8)
      • Relevant literature review
      • Preliminary data
      • Conceptual or empirical model
      • Justification of approach or novel methods
   D. Research plan (Chapter 9)
      • Overview of research design
      • Objectives, hypotheses, and methods (Chapters 7 and 9)
      • Analysis and expected results (Chapters 9 and 10)
      • Timetable (Chapter 11)
II. References Cited (Chapter 12)

Pitfalls

Perhaps the most common conceptual pitfalls are the failure to establish the general significance of your work or to link it logically to your specific project. Another typical error is devoting too much text to complex details or to your past accomplishments. Unless these are pertinent to your study, you could lose the at-

tention of reviewers. Other widespread weaknesses include a failure to construct testable hypotheses or identifiable aims; the construction of too many hypotheses
or goals; bad analytical or statistical methods; poor experimental design; weak questions; good big picture
but inappropriate tests for that question; too ambitious
a project for the time and money requested; inadequate
skills or credentials for the task proposed. Finally, there
are a number of procedural pitfalls that are to be
avoided at all costs. Most of these are obvious but can
be critical. For example, avoid alienating the reviewers
by permitting typographical errors, erroneous references, or incorrect or inconsistent numbers to creep
into the text. Follow all page-length guidelines. Present
a pleasant-looking document that is legible and logical
and, whenever possible, reader friendly. Again, evaluating other proposals and looking at your own project
with these specific issues in mind is of great benefit
throughout the writing process.

Developing Your Conceptual Framework and Significance Statement

Scientific proposals are always judged by their perceived significance. This is true whether you are writing for the NSF, the American Heart Association, a local conservation society, or a dissertation committee. Everyone who funds or supervises research inevitably asks what makes the proposed research "significant." If you cannot answer this question, stop writing and keep thinking.

All of the scientists we asked agreed that time spent early developing a proposal's significance, objectives, and hypotheses, aims or questions is time well spent. Remember that persuasive questions are essential for successful proposals.

Four cornerstones underlying good research are:

- Important questions

- Best and most appropriate methods or approaches

- Appropriate analysis and application of results

- Synthesis and timely dissemination of results

In this chapter we provide suggestions for conceptualizing and developing the first of these cornerstones, and we offer simple guidelines for writing a compelling significance section and placing it strategically in your research proposal.

Developing Your Significance Statement

The questions to be addressed by the study generally are featured with their justification in a significance section. Many scientists feel that this is the most important piece of a research proposal. A well-written significance section highlights the fundamental value of the proposed research, so many authors start the research unit with a significance section (see also Chapter 3). This section should be linked to the specific objectives, aims, questions, or hypotheses of your study (discussed in Chapter 7), which should follow closely in the proposal. The reviewer must find that the logic in this section is sound, that the ideas are exciting, and that the scope is reasonable within the time and budget you propose. Obviously this is not a trivial job.

To refine your thinking on the significance of your research, step outside your own discipline and immediate needs and take a broad and long-term view of your research. This perspective is essential for building a valuable and wide-reaching set of hypotheses. The goal is to end up with a pithy and accurate statement of the significance before you write the section. While writing or evaluating the significance section of a proposal we suggest that you:

- Look at the project from both a broad and narrow disciplinary view.

- Ask what scientists inside versus outside the field would perceive as the greatest contribution of this research.

- Consider both the empirical and theoretical contributions that may result from the study.

- Identify and contrast basic and applied uses of the data.

• Ask how you most expect and hope your research will be used by others.

• Compare contributions of the project that are likely to be important one year versus ten years after the completion of the project. Remember that the significance of a project changes with time and technology.

• Be your own best critic and ask how an impartial reader would dispute the claims that you have made.

Exercises for Developing Your Significance Statement

The following exercises should help you conceptualize and articulate your research proposals before you begin to write. View these drills as the building blocks for writing the entire proposal, and follow the five precepts for effective communication (organize, highlight, funnel, focus, and unify) while doing them.

EXERCISE 1. Prepare a ten- to fifteen-minute oral presentation or an outline of the conceptual framework for your proposal. Restrict yourself to the use of three slides or two pages. The challenge is to present the conceptual framework without reference to any specific system. For example, suppose you plan to investigate synergistic effects of exposure to toxic metals on reproduction and growth of oysters in estuaries off the Maine coast. For this exercise, distill the key conceptual points of your proposal: those ideas that are of importance beyond oysters, the specific metals you'll study, and estuaries in Maine. For example, you could focus on understanding synergisms among contaminants that occur in combination. The point of departure for your talk could be the need to tease apart mechanisms of interactions among contaminants in order to devise remediation strategies.

This activity forces you to articulate your general research question in broad terms, and to relate your study to the theoretical and empirical research that precedes it. If you must use a system to illustrate aspects of the discussion, do not use the system you intend to study.

EXERCISE 2. Distill your previous presentation to a five- to ten-minute oral presentation or brief written document of the significance and broad objectives of your research. This time you can refer to the specific system, cells, or organisms that you actually plan to study. When you can do this effectively you are probably ready to write a project summary and a brief yet pointed introductory significance statement for your proposal (see Chapter 6).

EXERCISE 3. Prepare a ten- to fifteen-minute presentation of the conceptual framework for your research project, focusing on the underlying quantitative, theoretical, and functional relations. Proposals often include or require a model or series of models to identify the key relations between processes (see Chapter 8). As in exercises 1 and 2, emphasize the relation between your study and the theoretical and empirical research preceding it. A graphic presentation of the conceptual or quantitative relationships can be particularly effective. Successful completion of this exercise can produce a piece to be used in the introduction and justification of your study.

EXERCISE 4. Identify a system (e.g., biochemical process, species, habitat) that is analogous to the one you plan to study. Give another five-minute oral presentation (or prepare a one-page written summary) on the significance and broad objectives of your research, this time structured entirely around the comparable system. This exercise forces you to consider the relations between significance and specific objectives more precisely, because sometimes objectives may not be as generally applicable as believed. This may make you

reevaluate your objectives with respect to your chosen
study system.

Crafting the
Significance Statement

An effective, engaging significance section mo-
tivates the reader to give your proposal a thorough appraisal, and it establishes the framework for the rest of
the study. The overall goals and significance should
also target information necessary in the background
section (see Chapter 10) and lead the reviewer directly
to the objectives and hypotheses or specific research
questions. If the significance section is not consistent
with the other sections, your proposal will not be persuasive.

A formulaic approach is rarely wise, yet we
agree with the widely held philosophy that an effective
significance section (generally one to two pages) begins
with the "big picture" motivating your work, elaborates
on the scientific context for your study, describes
briefly your own research plans, and restates the overall goals and expected results. Here are several tips for
producing your significance section, compiled from
our reviews of a number of excellent proposals.

- *Feature the significance section at the start.* This allows you to set the tone for the entire proposal. Some people place the significance section at the end of the proposal, but we find that this is much less effective. By that point most reviewers will have already formed a firm opinion of the study.

- *Keep the section short.* Don't dilute your message with detail, but be sure to elaborate beyond the project summary (see Chapter 6).

- *Funnel the reader.* Take the reader from your broadest goals to your specific aims. The more effective your funnel, the clearer the section will be. If you can write a statement in which your research appears to be the most logical and innovative approach to answering the question raised in the first or second sentence, you will have accomplished a great deal.

- *Explain the value of your work.* It is essential that you explain the value of your research questions in a manner that is accessible and convincing to scientists both in and out of your immediate discipline. Perhaps you have identified a glaring gap in knowledge. If so, explain what information is missing and describe how finding that information may lead to other important research. Maybe you plan to work on a process that has been identified in one system but not tested in other systems. You must convince the reader that applying the process and associated ideas to another system is important in some fundamental way. After reviewing this section the reader should understand how the successful completion of your work will advance the state of science in your field.

• *Link with other fields.* Successful research usually has significance beyond its immediate domain. Briefly explain the implications of your work for other fields, and how it can be applied in those fields. This makes your work more appealing, and it emphasizes that your study has breadth. It also provides a chance for you to identify and describe the broader societal impacts of your research.

• *Don't go overboard.* One important note of caution: be sure not to overreach. Reviewers become annoyed if the claims for significance are out of proportion to the specifics of the research.

You can accomplish your task in a variety of ways.

When composing the significance section, people often go back and forth between significance, objectives, and methods. We usually first write a rough draft of the significance section, develop the objectives and hypotheses, consider the methods, and then reconsider and rewrite the significance section. When writing this section, remind yourself of some often-encountered pitfalls: language that is too vague, use of overblown or naive statements of significance, repetition of other sections without additional detail, and confusing or jargon-heavy language that fails to engage the reader. As we stated earlier, it is usually most effective to delay writing until you have read a number of proposals in your discipline and completed the thinking for this section.

More Exercises for Writing the Significance Section

To help hone your skills, we suggest that you critique the significance paragraphs of other proposals. We do this in the classroom, using a supply of proposals given to us by their authors. With our students we evaluate the effectiveness of significance sections based on content, perceived importance of the questions, placement in the proposal, basic writing skills, and style. This activity is valuable at any time in the development of a proposal.

Experiment with names for this section. Different titles are commonly used. Some are more forceful than others, depending on the type and objectives of the proposal. Here are some examples from successful proposals. Which do you prefer?

• Overall Objectives

• Overview and Significance

• Significance and Project Objectives

• Statement of the Problem

Critique other significance sections. Consider the following excerpts from three significance sections and one entire significance section. Are they cogent? Are you engaged? Do they successfully cover some of the points listed above? Are the primary pitfalls avoided?

Field Measurements of Phytochelatins in Crops and Ecosystems Contaminated by Metals

Understanding how metal pollutants affect crops and forests is obviously of great importance to U.S. agriculture. Much research is aimed at elucidating the mechanisms of plant-metal interactions, including the induction of phytochelatins.

Source: F. M. M. Morel, excerpted from the Rationale and Significance section

The Evolution of Mate Choice in Damselflies

The broad objectives of our work are to explore the consequences of various speciation mechanisms to the assembly of real ecological communities over evolutionary time. . . . By understanding the role of sexual selection in generating new species, this system could become an exemplar for how many ecologically similar species can be rapidly introduced into an ecological system, and thus be a model for exploring how niche and drift-based mechanisms interact to shape the assembly and dynamics of real communities on a continental scale.

Source: M. A. McPeek and H. Farid, excerpted from the Objectives and Significance section

Understanding Hydraulic Conductivity in Aquifers from Above-Ground Measurements

Understanding hydraulic conductivity of a variety of aquifers has global importance and cannot be undervalued given the enormity of environmental problems related to contamination of aquifers. Our proposed technique, if successful, has the potential to revolutionize the way that groundwater studies are conducted. It may also greatly affect the rates of species extinction, global warming, and frequency of El Niño events.

Source: Excerpted from the Significance section of a fictional proposal

Understanding the Origin and Asymmetry of Life

This proposal addresses some of the key questions that must be answered if we are to truly understand life's origins here or anywhere else. It is designed to fully integrate our present and near-future observational and experimental knowledge with cutting-edge computer simulation techniques. It will serve as a springboard to future NASA missions.

Source: M. Gleiser, excerpted from the Relevance and Impact section of proposed research

Do you think that the relative weights given to justification of the big picture and its link to the aims of the study are effective in the following example?

Testing for Cascading Effects of Habitat Fragmentation

The results of this research will lead to a much richer understanding of human effects on the coastal sage scrub ecosystem. The proposed research capitalizes on the ongoing research of the investigators to generate a synthetic picture of ecosystem dynamics. It will also generate patterns and mechanistic hypotheses that will inform manipulative experiments that the investigators will describe in a forthcoming proposal to NSF. Furthermore, this research will be done in urban nature reserves in a region that has been highlighted for its importance as a test case for resolving human-nature conflicts in urban settings (reference given here). The proposed research will provide a diverse array of data that will lead to an in-depth mechanistic understanding that will inform reserve management in this region. The research also addresses fundamental ecological questions of the importance of top-down and bottom-up limitation and regulation of ecosystems (reference given here). These results will be directly communicated to reserve managers as well as to a wider scientific community through consultation with managers, participation in local and national meetings, publications in national and international journals, teaching undergraduate and graduate courses in conservation biology, and the participation of undergraduates, graduate students, and postdocs in the research.

Source: Modified from a proposal by D. Bolger and colleagues

Write your own significance section. After you have written a significance section, ask friends and colleagues to read it and tell you whether they think you've addressed the major points. Ask them if you followed the five precepts of organize, highlight, funnel, focus, and unify. In editing your significance section, keep in mind the essential points to cover and the pitfalls to avoid. If you have a solid, persuasive significance section, you are well on the way to completing a successful proposal.

A Title May Be More Important Than You Think

Titles are often written at the last minute and typically receive less thought than the rest of the proposal. But the title introduces your reader to the framework and perspective of the document. An effective title will capture that reader's attention and prepare him or her for the focus you wish to establish. The role of the title can be significant during the evaluation process, in which a review committee may collectively assess up to two hundred proposals. In some cases, members of a review committee may start by reading the applications with the most intriguing titles. In group discussions, a fragment of the title may become the way the proposal is referred to or remembered. When you can write a succinct and unambiguous title that captures the most important features of the work, you are ready to write a stronger, more focused proposal.

Components of a Good Title

The title must encompass the focus or concept of your proposal. If it is too descriptive, it may appear narrow, but if it is too broad, it may appear unachievable. An effective title accurately represents the content of the proposal. Practice is the soundest way to learn how to write a title. Read an assortment of titles, and as you read them, refine your own.

- Present your title in a clear, concise, meaningful manner.

- Avoid jargon and overstatement.

- Consider the impact of using buzzwords. Be aware that such language alienates some readers just as it attracts others.

- Avoid titles that are "cute" or too informal. This is arguably a matter of style, but we prefer titles that leave out the humor.

Exercises for Writing Titles

Ask yourself how your title can be clarified, shortened, and made more precise. Work on your title, share it, and rework it.

Categorize and modify existing titles. Titles come in many forms—questions, descriptions, bold statements. Each form works well in specific situations. Consider the following title: "Bedrock Influence on Soil Chemistry in Western Vermont." This descriptive title gives the reader a fairly good idea of the general topic but does not give any details on the system being studied or the questions being asked. "The Influence of Limestone on Base Saturation in Soils of the Lake Champlain Valley" is more specific, but its importance may not be immediately clear to nonspecialists. Both titles are more informative than "The Relationship Between Soils and Parent Material." Which of the first two titles is more desirable may depend on the content of the proposal, the agency or program to which it is being submitted, and the type of reviewers who are likely to read the proposal (specialists or generalists).

We contrived the following list of titles to represent the variety seen in proposals. Some are modified from published titles. Remember, evaluations are not necessarily conducted by specialists in the same field as the author of the proposal, so titles that can be understood only by a scientist in a particular discipline may not be very effective. Most of these titles can be improved. Are they understandable? Is it easy to guess the content of the proposal based on the title? Would a few word changes or a different approach result in a title that would be more interesting or effective to a nonspecialist in this discipline?

1. "Models of Impulsive Behavior in Mice"

2. "Temperature and Moisture: Controls on Global Carbon Cycles"

3. "Mathematical Modeling of Non-Linear Systems"

4. "Mechanisms of Cell Division and Differentiation"

5. "Large Mammal Response to Habitat Fragmentation: Reproduction and Survival Rates on Two Continents"

6. "Nutrient Cycling in Freshwater Ponds: The Role of Two Fish Species and Three Algae Species in Four Lakes"

7. "Hormone-Neurotransmitter Interactions in the Brain"

8. "Basic Research for the Future: Are There Enough Resources to Support the Human Population?"

9. "Erosion in Streams—Slip Sliding Away?"

10. "Socioeconomic and Environmental Drivers of Infectious Diseases in a Warmer World"

11. "Infrastructure for Studies of Planetary Formation Using the NASA Telescope Facility"

12. "Source and Fate of Particles in Wastewater Treatment Facilities"

13. "Acidic Permian Lakes: Understanding the Geochemistry of Ancient Acid Systems"

14. "Trade Policy, Child Labor, and Schooling in South Africa and Lesotho"

Construct titles from existing project summaries. Summaries (or abstracts) from proposals that have been funded by such government agencies as NSF and NIH are accessible on a variety of Web sites. One suggestion is to take agency-posted entries from the Web and evaluate both the summaries and the titles. Do the titles capture your attention? Do they encapsulate the material outlined in the abstract? You can also write your own titles for these proposals based on the summary.

Here are summaries from two funded proposals. Try to write a good title for each. Our students actually drafted titles very close to the originals, which underscores the clarity of these summaries.

Sample 1

The objective of the proposed research is to experimentally investigate the effects of temporal variation in resource supply on the outcome and dynamics of competition between consumers. The proposed research would use planktonic rotifers (small, multicellular zooplankton) as model systems. Experiments would test the predictions that temporal variation in resource supply changes competitive outcome, slows the rate of competitive exclusion, and allows competing species to coexist. These experiments would go beyond existing experimental studies by combining the following aspects: (1) using multicellular organisms instead of microbes, (2) using a temporal pattern of resource supply that is more realistic than that used in previous experiments, (3) measuring the effect of temporal variation in resource supply on the threshold resource concentration for positive population growth, and (4) predicting changes in competitive outcome, dynamics, and species diversity at different scales of temporal variation.

Source: K. L. Kirk

Sample 2

Overpumping of California's Salinas River Valley aquifers has prompted sea water intrusion, adversely affecting groundwater quality. Consequently, the sustainability of Monterey County agriculture is jeopardized. Growers recognize that the solution to preventing further intrusion lies in regulation of the aquifer, but they differ in their commitment to accepting groundwater upper pumping limits or pumping taxes.

In order to motivate grower effort to manage the groundwater resources more effectively, the effect of one grower's pumping on the water quality

and quantity of all other growers should be quanti-
fied. Because sea water intrudes the aquifer through
diffusion, growers who pump closer to the intruded
area cause more damage than those who pump at
the other end of the aquifer, given equal rates of
pumping across growers. This suggests that a policy
that varies by region would be more effective than a
basin-wide policy. We will simulate present conditions in the aquifer as well as impacts of policy alternatives through the use of a Geographical Information System (GIS) computer program.

Source: D. D. Parker

Here are some titles modified slightly from successful proposals. The actual titles from the two summaries presented above are at the end of this list. Notice that almost all of these titles are concise and clear.

1. "Analysis of Pesticide Transport Pathways and
   Degradation in Natural Wetlands"

2. "The Role of Ecological Interactions in Diver-
   sification: Phylogeny and Population Differentiation of Goldenrod in Two Communities"

3. "Human Modification of Landscape Function
   in New England and Florida"

4. "Predicting the Response of Terrestrial
   Ecosystems to Elevated CO_2 and Climatic
   Change"

5. "Thermal Conductivity in Oceanic Waters: Internal and External Factors"

6. "Patterns and Processes of Geomorphic and Hydraulic Adjustments During Stream Channel Recovery"

7. "Mercury Flux Estimates from Sites to Regions: Scaling-Up Across the Northern Hemisphere"

8. "Role of Plant Transcriptional Adaptors in Heat Shock—Regulated Gene Expression"

9. "Dynamics of the Mesosphere and Lower Thermosphere of the Arctic and Antarctic"

10. "Effects of Plasma Ionization on the Nonlinear Dynamics of Emission Spectrophotometers"

11. "Detailed Dynamics of Atmospheric Photo-Reactions"

12. "Reconciling Molecular and Fossil Evidence on the Age of Angiosperms"

13. "A Study of the Abundance and $^{13}C/^{12}C$ Ratio of Atmospheric Carbon Dioxide and Oceanic Carbon in Relation to the Global Carbon Cycle"

14. "The Role of Identified Cells in Directional Motor Behavior"

15. "Genetics, Mechanism, and Regulation of Protein Synthesis in *G. hypothetica*"

16. "Segment-Based Acoustic Models for Continuous Pattern Recognition"

17. "Resource Competition Between Rotifers in a Variable Environment"

18. "Spatially Efficient Management of a Sea Water—Intruded Aquifer"

Write and critique your own title. Some people write a title as they begin their proposal, and others wait until the proposal is completed. Reviewers often say that they can tell new proposal writers by the titles of their proposals, which tend to be wordy and either highly specific or overblown. Try a number of titles, and experiment with the various forms. Feedback from friends and colleagues is extremely useful. Another approach is to ask yourself how you would want people to refer to your proposed project. Try to construct a four- to eight-word descriptor of the project and then expand it slightly only if absolutely necessary.

One of the first exercises we do with our class is to ask the students to write their titles on the board. As the group discusses the titles, they are usually changed, often quite radically, becoming more precise and focused. Students who have the most difficulty with this exercise are usually less certain about the direction of their work, so this can be helpful as a diagnostic exercise.

The Project Summary Guides the Reader

As the first and shortest section in an NSF-type proposal, the project summary serves several vital functions. It is where you frame the goals and scope of your study, briefly describe the methods, and present the hypotheses and expected results or outputs. The project summary (some people use this term synonymously with "abstract") is the initial description of the project seen by reviewers. A convincing and exciting summary captures their attention and interest, and it establishes a strong tone for the entire document. It is critical to set up the proper expectations, to avoid misleading readers into thinking the proposal addresses anything other than the actual research topic. What a challenge—to be clear, concise, accurate, and exciting, all in fewer than three hundred words.

When assessing a proposal, reviewers use the

summary as a template or guide to the document. Their impressions of the summary are critical. Program directors frequently rely on the summary when choosing ad hoc reviewers. Summaries also are used later to remind evaluators of the key elements in the design and of the expected outputs. This role is particularly important for proposals being formally evaluated, and it may be referred to repeatedly during review panel discussions. Panelists judge many proposals in a short, intense time period, but you can make your work stand out by providing concise, precise, memorable sentences and phrases in the project summary or goals section. Remember that whatever you write in the summary will be used to highlight major aspects of your study, so be sure to state exactly what you mean.

There is no one template for designing an effective project summary. Some of the most compelling summaries start with a broad statement of purpose and then funnel the reader to the specifics of the proposed work (see Chapter 3). Some authors write the summary and then use it as an outline for the grant application; others write it after the rest of the proposal is completed. The essential feature is that the summary must accurately encapsulate the most important elements of the project. Summaries from several funded proposals appear below (with the permission of their authors). You may wish to critique these before writing your own project summary.

Elements of Effective Project Summaries and Specific Aims

Our discussion centers on a model for NSF basic research proposals in which the initial section is termed Project Summary. For NIH proposals the Abstract page is a brief description of the Research Plan; the Specific Aims section is a more specific summary of long-term objectives, research objectives, and any hypotheses (Reif-Lehrer 2005); every agency may have a different title or slightly different format for the precise part of the document where the proposed research is summarized. However, the goals of these sections are generally similar. Before writing this element, be sure to obtain precise specifications.

The NSF's description of the Project Summary reads: "[This] should not be an abstract of the proposal, but rather a self-contained description of the activity that would result if the proposal were funded. . . . [It] should include a statement of objectives, methods to be ployed and the significance of the proposed

vity. . . . Insofar as possible, [it should be] understandable to a scientifically or technically literate lay reader."

This general statement distinguishes a proposal summary from a manuscript abstract. In a proposal for a research project, you report what you plan to do and stress why your work will be influential, not what you have done and why it is important. Emphasis is placed on significance and context, and it is essential to establish that there will be consequential outputs and "broader impacts resulting from the proposed activity." The "broader impacts" product of the proposal is explicitly stated as one of two merit review criteria that must be included in the one-page Project Summary of NSF proposals. The GPG states in bold letters that proposals lacking a description of intellectual merit and the broader impacts will be returned without review.

For a task-oriented proposal you highlight the expected outputs and the novel aspects or particular qualifications that justify your selection over others for the specified goals. Results are not included in a summary, although you may refer to your previous research to make a point or to establish your ability to accomplish the proposed task. Length limits for summaries vary with the funding agency, but they rarely exceed a single-spaced page. Agencies expect scientists to write the summary in a way that can be understood by a scientifically literate lay audience. This can be difficult for some writers, but it is worth taking the time to write this brief but important section in a way that will make your proposal topic more understandable for nonspecialists and the general public.

Before you begin to write your summary, be aware of the conventions in your field (e.g., are hypotheses, specific aims, or objectives generally presented in the summary?). Ask friends and colleagues for copies of summaries they have written, and read or download summaries, specific aims, or abstracts of funded proposals (they are in the public domain) at the Web sites of major funding agencies; see also the examples below.

The two-paragraph project summary model. In your summary you have just a few sentences in which to direct the reader from the most general and broad significance of your proposed research to its specific details. A number of styles and formats can be successful. Some summaries begin with a bold statement: "The proposed work will test the hypothesis that . . . " Others employ a more gradual development of ideas or build chronologically from early views to the current state of the field. The second style is logical, but it may fail to capture the reader's interest at the start because it is less effective at highlighting important points. We encourage you to begin by writing a two-paragraph summary. Use the first paragraph to introduce the problem and describe the work, and the second paragraph to emphasize the potential outcome and significance, including the broader impacts.

Paragraph 1

- Develop the broadest context for the research in the first one or two sentences.

- State your research questions as testable hypotheses or, where appropriate, as objectives. Never propose untestable hypotheses or questions or goals that cannot be met with the proposed research.

- Identify gaps in current knowledge and state how your questions fill those gaps or lead the field forward. Establish the overall importance or relevance of your work. This tactic will also help to justify funding your study relative to other well-conceived studies.

- If appropriate, include preliminary results of your own work; these make further work compelling and establish your credibility.

- In the last few sentences of the paragraph, give a detailed, succinct description of the actual work that you will do.

Paragraph 2

- Briefly summarize or describe techniques, study sites, and, if appropriate, the taxonomic names of study organisms (this may be in the first or second paragraph).

- Discuss the projected results or output from your proposed study.

- State how your work will advance your area of study; perhaps include a phrase or sentence on the implications of your work for other fields or issues. Be careful about making statements that cannot be supported by your work.

- Specify the broader significance of the work, including how your proposal will advance discovery and understanding while promoting teaching, training, and learning or providing other broad societal benefits.

The following project summary illustrates some of these ideas: it uses the two-paragraph model, and it provides a context for the study. There are testable hypotheses stated clearly, and the broader value of the study is stressed.

Role of Winter Water Relations in Determining the Upper Elevational Limits of Three New England Conifers

Winter desiccation is recognized as an important stress factor in coniferous forests, and it may limit conifer distribution. Most research to date has focused on desiccation at alpine treeline, whereas little attention has been given to its role in establishing the upper elevational limit of low-elevation conifers. Our objective is to test the hypothesis that winter water relations limit the upper elevational range of low-elevation evergreen conifers in New England. This will be the first study to examine desiccation stress in non-subalpine conifers. The winter water relations of three low-elevation conifers will be examined: white pine (*Pinus strobus* L.), eastern hemlock (*Tsuga canadensis* [L.] Carr.), and red pine (*P. resinosa* Ait.). Each of these three species differs in its habitat preference and growth strategy. Preliminary results indicate that older foliage in each species can reach water levels expected to cause desiccation damage. Our approach will use physiological measurements of trees (relative water content, water potential, and cuticular resistance) collected near the upper elevational limit of each species during the winter to assess desiccation stress. These data, along with micrometeorological data collected at field sites, will be used to predict winter water relations. We will test the following hypotheses: (1) water levels in foliage near the upper elevational distribution of each species will approach or fall below lethal desiccation levels; and (2) cuticular resistance will decrease over the course of the win-

ter. Even if this work does not support these hypotheses, the understanding of conifer responses to winter climate will be greatly increased.
This study will be of value to plant-stress physiologists and plant ecologists. It is unique in that it will combine field assessments of desiccation with micrometeorological measurements in a model, allowing plant-water relations to be explicitly coupled to climate. Such an approach sets the stage for future studies of limitations by winter desiccation, using other species and under conditions imposed by a changing climate. Broader impacts of this work: through the use of undergraduate field assistants, we will integrate field science into the education of diverse audiences, furthering their understanding of biological and environmental processes and helping to train young scientists.

Source: Modified after R. L. Boyce and A. J. Friedland

Examples of project summaries on the Web. While writing and revising this chapter we read hundreds of project summaries from grant applications for projects that were funded by U.S. agencies. The summaries were from a variety of fields represented in this book (e.g., environmental science, ecology, molecular biology, earth sciences, atmospheric sciences, and neurobiology) and available on the Web or through published annual reports of major funding agencies. We focused mainly on summaries from NSF and U.S. Department of Agriculture. Summaries from these proposals varied greatly, yet we found many similarities, which we incorporated into the suggestions presented in this chapter. For example, all authors specified their research questions and explained the significance of their research topic. By reading these summaries and others you will decide which styles and approaches you prefer. (See the list of Web addresses at the end of this chapter.)

Here is another summary that combines many of the important features we have discussed.

Neural Network Model for Chemotaxis in C. elegans

This research addresses the question of how the brain uses sensory information to select the most effective behavior in a given situation. This question is addressed by studying the nematode worm *C. elegans,* an experimental organism whose compact nervous system of only 302 neurons is unusually well suited to investigating relationships between brain activity and behavior. The main focus of the research is to build and test a computer model of the nematode's neural network for chemotaxis, a simple yet widespread form of spatial orientation behavior in which an animal finds food, shelter, or a mating partner by directing its locomotion toward the source of an odor or taste. The model will be used to test the idea that the nematode's chemotaxis network utilizes separate neuronal pathways to signal increases and decreases in sensory input, much like the visual system in higher organisms, including humans, and should provide new insights into how neural networks function to control adaptive behaviors. Broader impacts: Nematode chemotaxis is accessible to young scientists and the lay public alike. The PI hosts visiting undergraduates and high school students from underrepresented groups. University undergraduates have participated in this project from its beginning and will continue to do so. These researchers (currently two honors college students and two freshmen) do real science, including laser ablation of neurons and quantitative assessment of behavior. Simplified versions of the models developed in NSF-sponsored research are a prominent unit of the PI's course in computational neuroscience.

Source: S. Lockery

Exercises for Writing a Project Summary

Before you write your own project summary, read and critique a number of others. For example, modify a few sentences in our previous sample summaries (Chapter 5) to see whether you can improve them. Try adding hypotheses or working with the placement of the significance statements. Ask yourself: Is the writing easily understood by the reader? Does it flow from topic to topic? Are transitions smooth and clear? Read the summary out loud to see.

Critiquing other summaries. We have constructed

two versions of a project summary. See how you can improve on each version.

Version 1

Effects of Nutrient Additions on Red Spruce Health and Nutrition

The purpose of the proposed study is to determine the effects of nutrient additions on carbon fixation and foliar nutrition of high-elevation red spruce in the northeastern United States. Trees in this ecosystem are declining, a circumstance that has been attributed in part to a changing chemical environment. Earlier work has shown that nitrogen and sulfur inputs are quite high, and nitrogen saturation has been suggested as a cause of decline. Work from the southern Appalachians suggests that acidic deposition—induced calcium deficiency, perhaps coupled with aluminum mobilization, causes increased rates of respiration and reduced photosynthesis: respiration ratios which lead to reduced growth. Work with potted spruce seedlings from the South has confirmed this, and it is consistent with patterns observed in the field. Preliminary data by our group suggest that in New Hampshire, 1) spruce respond positively to additions of nitrogen, and 2) there are adequate supplies of base cations. Data from New York suggest that 1) nitrogen additions reduce foliar growth, and 2) calcium is limiting. We recognize that some of these findings are, in part, inconsistent with previous findings, and thus we wish to extend our investigations.

We propose to conduct a field study using naturally grown spruce saplings on Mount Jefferson (White Mountains, New Hampshire, United States) and Mount Marcy (Adirondacks, New York, United States). Fertilizer treatments using N, Ca, and/or Mg will be applied over the three-year course of the study. Photosynthesis and dark respiration will be measured throughout the growing season to determine the response to different treatments. Foliar concentrations and contents of N, base cations, and Al will be analyzed to determine their effects on carbon fixation rates.

This study will increase our understanding of the impact of global changes in atmospheric chemistry on high-elevation red spruce—balsam fir forests. However, our research has potential significance beyond this particular ecosystem, for it will show how conifers in general respond to abiotic stress. Chronic levels of stress, such as those induced by global change, often initiate forest declines. The early stages of decline are often subtle and therefore overlooked, creating difficulty in identifying the onset of decline. The research proposed here will make use of a species that is known to be in decline, across a gradient from low to high levels of decline.

Source: A. J. Friedland

Version 2

Effects of Nutrient Additions on Red Spruce Health and Nutrition

Nutrient deficiencies and imbalances are known to cause problems in plant growth and metabolism. High levels of nitrogen and sulfur deposition in the northeastern United States are suspected of causing nutrient imbalances at higher elevations. Many hypotheses related to N, S, Ca, and Al have been offered to explain the decline of red spruce at high elevations in the Northeast. Preliminary data by our group suggest that in New Hampshire, 1) spruce respond positively to additions of N, and 2) there are adequate supplies of base cations. Others have suggested that Ca supplies may be limiting. We propose to conduct a field study using naturally grown spruce saplings on Mount Jefferson in the White Mountains of New Hampshire. Fertilizer treatments of N, Ca, and/or Mg will be applied over the three-year course of the study. Photosynthesis and dark respiration will be measured throughout the growing season to determine the response to different treatments. Foliar concentrations and contents of N, base cations, and Al will be analyzed to determine their effects on carbon fixation rates.

This study will increase our understanding of the impact of atmospheric deposition of pollutants on high-elevation red spruce—balsam fir forests, and it may provide information on how conifers in general respond to abiotic stress. Chronic levels of stress can initiate forest declines in other temperate coniferous forests.

Source: A. J. Friedland

We feel that the first version could be substantially improved. For example, findings are reported in a chronological or sequential order. "Work from . . . " and "Work with . . . " constructions become repetitive and are not the most integrative and synthetic way to present information. The questions and hypotheses are not explicitly stated. Do you feel that the potential significance of the work is clearly stated? One way to amend this abstract would be to delete most of the first paragraph, describe the general area of nutrient deficiencies first, and identify the unanswered questions more succinctly. The second version is better but still fails to articulate hypotheses.

Here are some other summaries from successful proposals that we obtained directly from the authors or downloaded from agency Web pages (used with permission of the authors). All are quite forceful, but even the best summaries can be improved. Try to enrich each one.

Biochemistry of Fatty Acid Transport in Escherichia coli

In all organisms, fatty acids (FA) and their derivatives are components of membranes, are sources of metabolic energy, and are effector molecules that regulate metabolism. This research is on the transport of long-chain fatty acids (C14–C18) into the cell, followed by their enzymatic conversion to coenzyme A thioesters prior to metabolism. These FAs traverse the cell envelope of Escherichia coli by a specific, energy-dependent process that requires the outer membrane-bound FA binding protein FadL and the inner membrane associated acyl CoA synthetase (ACS). ACS activates FAs concomitant with transport and results in net FA accumulation in the cell against a concentration gradient. Processes that govern FadL-mediated long-chain FA transport across the outer membrane will be determined by i) evaluating the topology of FadL using limited proteolysis and protein modification and ii) defining the FA binding pocket within FadL using the affinity labeled long-chain fatty acid 9-p-azidophenoxy nonanoic acid (3H-APNA). The contribution of acyl CoA synthetase to long-chain FA transport will be evaluated by i) defining the ATP and FA binding domains within ACS using the affinity labeled ligands azido-(32P ATP and 3H-APNA, respectively), and ii) mutagenesis of the FadL gene at specific sites involved in CoA and/or FA binding. Studies are also being conducted to define protein-protein interactions between the membrane-bound (FadL) and soluble protein components of this transport system using Far Western analyses, and by performing experiments with glutathione S-transferase (GST) and histidine fusion proteins. Soluble protein components may interact with FadL. The H1/FA cotransporter in the inner cell membrane, and acyl CoA dehydrogenase and acyl CoA binding protein in the cell cytosol, may bind specifically with ACS.

Source: P. N. Black

Modified from "Assessment of the Severe Weather Environment Simulated by Global Climate Models"

Severe thunderstorms and tornados are very important mesoscale weather events in the central United States because of their high frequency and intensity in this region, and the damage and loss of life that they cause every year. Recently, it has been shown that the frequency of favorable conditions for significant severe thunderstorms and tornados can be estimated for the United States and other regions using global atmospheric reanalyses with spatial resolution on the order of 200 km and temporal resolution of 6 hours. Global climate models are unable to simulate severe thunderstorms and tornados because their spatial resolution is too coarse to be able to simulate such mesoscale events. However, they should be able to simulate the environmental conditions under which such severe weather develops, including abundant lower tropospheric moisture, steep mid-tropospheric lapse rates, and strong tropospheric wind shear. High space and time resolution data from control simulations with global climate models archived at NCAR will be used to estimate the frequency of favorable conditions for severe weather, as simulated by the models. The climatological distribution of the severe weather environment in the model simulations will be compared with that from the reanalyses, including the seasonal and geographical variations and its interannual variability. Outcomes from this research will include a detailed assessment of global climate model simulations of the environmental conditions determining severe weather. This will provide a better understanding of some of the causes of model problems with simulation of warm season intense continental convective. Broader impacts: In addition to the scientific outcomes, the graduate student employed on this project will gain valuable training and experience in climate diagnostics, climate modeling, and severe weather, which will allow him/her to contribute better to future research and development in climate change and its impacts in the United States

Source: D. Karoly and H. Brooks

Here are a few examples of specific aims:

From a proposal titled Behavioral and Physiological Responses to Anabolic-Androgenic Steroids

Specific Aim 1: To establish the dose-response characteristics and the role of the type of intruder on the expression of AAS-induced offensive aggression in male and female mice. Female and male mice will be tested in response to different types of intruders for offensive aggression and following treatment with six doses of a cocktail of AAS that vary in their androgenic and estrogenic properties and are commonly self-administered by humans.

Specific Aim 2: To determine if loss of andro-
gen receptor (AR) or estrogen receptor (ER) signaling induced by pharmacological inhibitors interferes
with AAS-induced aggression in either male or female mice. AAS can be metabolized in the central
nervous system to both androgenic and estrogenic
derivatives. The relative importance of AR versus ER
signaling in mediating AAS-induced aggression in either sex is not known. We hypothesize that both AR
and ER signaling will be involved in mediating AAS-
induced aggression, but that AR signaling will play a
more important role in female mice and ER signaling
will be more important in male mice. We will assess
offensive aggression in wildtype mice of both sexes
treated with a cocktail of individual AAS in conjunc-
tion with the AR antagonist, flutamide, or the ER antagonist, CI-628.

Source: A. Clark and L. P. Henderson

*Collaborative Research: The Mixed-Race House-
hold in Residential Space: Neighborhood Context,
Segregation, and Multiracial Identities, 1990–2000*

First, we intend to map and analyze the neighborhood geographies of mixed-race households in
1990 and 2000. This will provide answers to questions such as: Do the processes that generate and
sustain segregation, as we currently understand
them, apply to mixed-race households? Do mixed-
race households live in segregated neighborhoods or
diverse neighborhoods? And how does this vary
with the race, class, or nativity of the partners?

The second aim turns the first inside out: instead of assessing the influence of segregated
neighborhoods on the geographies of mixed-race
households, this phase of the research drives at how
much racial mixing within households contributes to

racial mixing within neighborhoods. Segregation measures typically rely on counts of individuals in neighborhoods and ignore the mixing of groups in households. This part of the research asks how much neighborhood diversity is accounted for by mixing in households. And how will rising rates of mixed partnering affect change in levels of neighborhood scale segregation?

Source: M. Ellis, S. Holloway, and R. Wright

Search the Web. The following web addresses, accurate as of press time, provide access to summaries of funded grants from federal agencies. These sites also provide much more information about suggestions and requirements regarding proposals:

- National Science Foundation:
  http://www.nsf.gov/

- USDA National Research Initiative Competitive Grants Program:
  http://www.csrees.usda.gov

- Environmental Protection Agency:
  http://www.epa.gov/docs/ord/

- National Institutes of Health:
  http://www.nih.gov/
  and
  http://crisp.cit.nih.gov

Objectives, Hypotheses, and Specific Aims: An Exhaustive List Is Exhausting

A beautifully crafted document, or a convincing and exciting significance statement, means little if the research objectives and tests are ill conceived, poorly stated, or absent. Our class concentrates heavily on constructing, deconstructing, and reconstructing each others' objectives, hypotheses, and specific aims. If working alone, be sure to find a few colleagues who are willing to exchange ideas. Reviewing your hypotheses and objectives with others may be one of the most beneficial activities you can pursue.

Development of hypotheses or aims generally precedes proposal writing. Most readers of this book probably have already identified a series of objectives and associated hypotheses. Solicit feedback prior to writing by presenting these hypotheses to colleagues and mentors to determine whether they are rigorous, testable, and engaging. The goal now is to articulate those ideas in writing and to place them into a proposal. A key aspect of this process is making sure that the hypotheses are consistent with the significance statement and linked properly to the objectives.

Objectives, Hypotheses, and Specific Aims

Objectives usually refer to broad, scientifically far-reaching aspects of a study, and sometimes they verge on significance statements. (In some fields, goals are synonymous with objectives.) If both significance statements and objectives are included, the objectives generally are more focused. Objectives can also pertain primarily to contributions or novel uses of the data within the scientific community. The following examples are modified from proposals that we have written or that have been shared with us by the authors.

Our objectives are:

- to further our understanding of the implications of global climate change in freshwater lake plankton communities.
- to lead to more informed policy decisions about the effect of electromagnetic radiation on humans residing near high-voltage power lines.

- to understand how cell division and differentiation are regulated by extracellular and intracellular signals.
- to evaluate mechanisms leading to species coexistence in marine intertidal communities and compare magnitudes and scales of effects.
- to provide the first complete data base for the assessment of toxic metals on reproduction.
- to develop an analytical framework for classifying brain potential analysis of motor function and decisions.

Hypotheses usually refer to an even more specific set of testable conjectures than do the objectives. A well-formulated hypothesis leads directly to the experiments and sampling programs that form the basis for the research. Keep the number of hypotheses reasonable: it is important to strike the proper balance between too many and too few. If you present too many, your proposal will confuse the reader and reduce its effectiveness. One of us wrote a proposal with twenty-seven hypotheses. Rather than being dazzled with the breadth and depth of thinking and synthesis involved in the study, the reviewers were baffled and bored. They also perceived a lack of focus—no surprise! After the proposal was rejected, the program director suggested reducing the number of hypotheses to five or fewer to focus the effort. We agree with that general recommendation, though there is no magic number for how many hypotheses to include.

We have adapted some hypotheses from our own and others' proposals. Is it apparent how these hypotheses differ from general significance statements or objectives?

- We hypothesize that lead is complexed by a chelating agent associated with adventitious roots, transported across the membrane, and stored in the inner cortex.

- Channel roughness is greater, and velocity, stream power, and shear stress are lower, in restored reaches versus unrestored reaches.

- Differences in temperature and humidity among sites persist across years, despite natural yearly variation in climate and other environmental variation.

- Zinc can displace other metals on enzyme-active sites.

- Mineral weathering in the lower soil horizons provides more than 80 percent of the cations lost from the ecosystem to stream water.

It is the convention to include a section describing the project's specific aims when writing proposals for NIH and some other agencies. Proposals tend to have two to four specific aims, which are often similar to hypotheses. Each specific aim generally focuses on a particular question or hypothesis and the methods needed and outputs expected to fulfill the aim. Frequently, the rest of the proposal will be organized around each of the aims. The most effective proposals often use the significance statements to unify the specific aims and to establish a progression or a funnel from one to the next. As for objectives and hypotheses, there should not be too many, and they should be concise and easily understood if they are to be successfully conveyed to the reader.

Here are some specific aims that we have modified from our own and others' proposals.

- We will determine the crystal structure of all six minerals in the source rock.

- High-resolution structure of a full-length dynamin protein: We will investigate the protein's global molecular mechanism and determine how it regulates itself.

- We will establish differences in temperature and humidity among sites over time.

- We will determine whether the organism promotes biofilm formation by secreting an inhibitory compound.

Linking Objectives and Hypotheses to the Significance

Although objectives, hypotheses, aims, and overall significance refer to different key features of a research proposal, they are tightly linked and must work well together. Each relies on the other for its validity and purpose. The significance statement is the most general and far-reaching description of the research; objectives are usually more focused than is the significance, and hypotheses or specific aims are more specific than are the overarching objectives. Objectives, hypotheses, and aims are more likely to identify particular processes, organisms, or locations than are significance statements.

To illustrate the relation between significance, objective, and hypothesis, we return to an example from Chapter 4 (from F. M. M. Morel). In this example, the overall significance statement introduces the topic of metal pollution and its major effects on agriculture as the overall focus of the study:

> Understanding how metal pollutants affect crops and forests is obviously of great importance to U.S. agriculture.

One of the objectives in that same proposal identifies the metal pollution of interest as coming from smelters and being airborne:

> [We wish to determine whether] in areas of high
> metal pollution, such as those near smelters, plants
> are exposed to metal stress through direct airborne
> pollution or [indirectly] through accumulation in
> soils.

Later on, the author presents this hypothesis, which
names the metals as Ni and Cu, the mechanism as atmospheric deposition, and the focal plant species as
paper birch:

> To investigate the hypothesis that current atmos-
> pheric deposition of nickel and copper aerosols is the
> dominant source of metal stress in vegetation sur-
> rounding Sudbury, Ontario, seedlings of *Betula pa-
> pyrifera* will be placed at each sampling location.

Note how the author becomes increasingly specific in
the move from significance to objectives to hypotheses,
and how each is closely linked to the others.

Here is another illustration of the progression.
The significance statement is broad in scope and re-
lates to a problem of international concern, namely
global climate change. In creating this example we de-
liberately avoided the popular terminology, but that is a
matter of personal preference:

> We wish to understand the biological implications of
> projected increases in global temperature on fish
> populations.

The objective is much more precise and is directed to a
particular type of system (salmonids on the rearing

grounds), yet it remains somewhat inclusive (not regionally focused, nor species specific):

> We will quantify responses of salmon to predicted increases in summer temperatures in their rearing grounds.

Finally, the two hypotheses that derive from this objective lead directly to an easily identifiable set of experiments or measurements. They are species-specific and address certain demographic traits and rates:

> A water temperature increase of 18C in May will advance the hatching date of Atlantic salmon by 2 weeks.

> Advancing the hatching date of Atlantic salmon by 2 weeks will reduce survival rates.

In this example we also have become increasingly specific, moving from significance to objectives to hypotheses.

Here are some examples in the format of specific aims rather than hypotheses.

- Characterize the effects of individual growth and metabolic rates in controlling the bioaccumulation of methyl mercury in fish.

- Determine the biological consequences of arsenic-driven alterations in steroid-receptor signaling in all cultures.

- Investigate the molecular-genetic alterations associated with chemical carcinogen exposure in humans to identify biomarkers of low-dose exposure.

Placement in the Proposal

Successful proposals often feature significance, objectives, and hypotheses sections near the start of the proposal, but there is no specified location for them. In contrast, specific aims in NIH proposals are usually the first section of the proposal. Authors usually introduce objectives and even hypotheses in the project summary or aims, and the objectives almost always appear in the significance section of proposals. You need to strike a balance between early presentation and appearing repetitious when you discuss the same material in greater detail in later sections. Many authors insert the hypotheses in a number of locations in a proposal, presenting greater detail with each mention.

In the following example, the general hypothesis is first stated in the title and then repeated with greater detail, and in a different fashion, in various sections of the proposal:

- *Title.* "The Role of Temporal Control Genes in Specifying the Timing of Events in the Nematode *C. elegans*"

- *Project summary.* "The broad goal of this work is to understand the genetic and molecular mechanisms of the temporal control of cell division and differentiation using the nematode *C. elegans* as a model."

- *Introduction and background.* "Animal development is a complex schedule of processes that are controlled by genetic and other factors."
- *Significance section.* "The *C. elegans* genes offer an opportunity to study the genetic and molecular mechanisms controlling cell division and differentiation, processes central to all multicellular development."
- *Research design and methods.* The hypotheses are presented in a preface to each set of experiments which are designed to test them.

Source: V. Ambros

Regardless of where you cite your objectives and hypotheses, and depending on the requirements of your funding agency or dissertation committee, using headings and subheadings highlights the importance of objectives and hypotheses within the proposal. It also makes it easier for the reviewers to find them. Be sure to follow the conventions of your field. For example, the practice in some fields is to use the traditional null hypothesis; that is, no matter what you think the outcome of your study might be, you state that there will be no effect (e.g., "The test drug will have no effect on the population."). Other fields are more ac-

cepting of positive hypotheses ("The test drug will reduce symptoms in more than 75 percent of the test
population."). The terminology used may also vary
across disciplines (e.g., hypotheses are sometimes
referred to as questions).

Exercises for Writing Objectives, Hypotheses, and Specific Aims

Our class uses the following exercise to formulate a very tight set of hypotheses or specific aims prior
to writing the accompanying text. This formulation
step may take several weeks as you review the critical
feedback on your objectives and hypotheses and then
revise and restate. This exercise is a strong follow-up to
exercises 1 through 4 in Chapter 4.

> **EXERCISE 5.** Prepare a ten- to fifteen-minute pre-
> sentation of the objectives and specific aims or
> hypotheses of your study, focusing on the direct
> links between them. As in the previous exer-
> cises, justify the importance of your hypotheses
> with respect to the larger field of theoretical and
> empirical research preceding it. Hypotheses
> must be concise and easily understood, and the

flow among the hypotheses must be logical. If you are writing this exercise, confine the text to fewer than two pages. Work to develop a sensible progression and transition among ideas.

We again urge you to collect as many proposals as possible and to evaluate them on your own or with a group. In addition to considering logic and flow, also assess the style of presentation. One widespread problem we see is that some authors number their objectives or hypotheses in a confusing manner (e.g., "I.A.b.iii"), which can diminish the strength of the final work. This is a minor point, but as we've noted elsewhere, a clear presentation of objectives and hypotheses is extremely important.

Lay the Foundation in the Introduction

Once you provide reviewers or your committee members with a perspective on the significance of your research and steer them toward your objectives or aims, your work begins in earnest. You now need to create the essential elements of what NSF terms "The Project Description" and NIH calls "The Research Plan."

The introduction or background is a major element of the project description or research plan. In this section you should review the literature and stress key references. Introduce relevant conceptual, theoretical, or empirical models, and discuss the need for new methods or technologies if they are pivotal to your research. To set the stage for your proposed study and to establish your ability to accomplish the task, a summary of your own relevant prior research or preliminary results is usually included here or in a separate section. The main goal of the introduction is to present the background material necessary to lead the reviewer to your objectives, hypotheses, and proposed research based on an overview of current research.

Suggested Order of Presentation of the Project Description

III. Project description (following from Chapter 3)
  A. Results from prior agency support (this chapter)
  B. Statement of the problem and significance (Chapter 4)
  C. Introduction and background (this chapter)
    • Relevant literature review (this chapter)
    • Preliminary data (this chapter)
    • Conceptual, empirical, or theoretical model (this chapter)
    • Justification of approach or novel methods (this chapter)

The introduction must funnel the reader from a general review of the literature to your specific study. After reading a persuasive introduction, the reader should exclaim, "Of course! What a great idea for a research project. Why didn't I think of that?" The introduction section provides an excellent opportunity to establish why your work is compelling and to capture the attention of a committee or funding agency.

The Basics of Introductions

Results from prior agency support. An introduction may begin with a section for prior results, which

generally refers to either 1) results of research that was
previously funded by the agency to which the current
proposal is being submitted, or 2) your own preliminary data that are essential to the proposed work. This
type of information may not be needed in most graduate dissertation proposals, in which preliminary results
may be more effectively blended directly into the introduction—background section. However, if you or any
of your co-investigators have prior research that has
been funded by the agency to which you are applying,
you may be required to summarize your results in a
specific section. NSF and some other agencies require
a feature called "Results from Prior Agency Support."
Typically this appears as the first part of the project description, preceding even the significance statement.

　　　Reviewers of proposals are always asked to comment on an applicant's productivity and on the quality
of his or her previous work, so this section may have an
important influence on whether a project is funded.
Keep your prior research statement concise, citing your
published work where appropriate. We advise using
this section to prove your past success and productivity
and highlight previous results that have been insightful and that lay the groundwork for the new research.
Seek guidance from the program director at your
agency for specifics on the types and amount of infor-

mation typically contained in this section. Remember that for most proposals, space is at a premium. Allotting too much text to this part of the proposal will naturally restrict the space that can be used to present other material. Avoid providing information that will distract from the proposal at hand.

Even if it is not required, a prior results section is highly recommended if the current proposal is an extension or continuation of previously funded work, or if a result or discovery from the prior work is relevant to the proposed work. If your prior work is not related to the proposed work but a statement is still required, keep the mention short and consider putting it at the end of the project description, where it is less likely to draw attention away from the current proposal.

Preliminary results. If you are a beginning researcher or are new to a particular field, you will probably not have results from previous related work. However, even new researchers may have generated preliminary data that should be included in the proposal. A prior or preliminary results section within the introduction may be an appropriate location for your own unpublished, preparatory data. By placing your preliminary results here you can establish your competency or the likely success or novelty of your proposed research. Avoid overstating the implications of your re-

sults, and avoid concentrating on them to the detri-
ment of making the case for your new project.

Provide a Strong Foundation
for Your Research in the Body
of the Introduction

As in the introduction to a manuscript, you
must cover the key concepts, previous work, and important publications that will allow an informed scientist from another field to understand the motivation for
your research. At the same time, this section should
not be too elementary for experts in the field. To para-
phrase Day and Gastel (2006) in their book on writing
scientific papers: An ideal proposal introduction allows
the reader to understand and evaluate the proposed
work without needing to refer to previous publications
on the topic.

There is no one right way to organize your introduction, but we continue to suggest funneling from
the general to the specific. Some authors build from
the specific to the general, however. Try articulating
your own arguments in each of these ways to see which
allows you to be the most concise, logical, and interest-

ing. Most of all, watch the length of this section. Inexperienced writers often include too much background material. The background should reinforce the need for the proposed work but avoid tangential material that dilutes the message.

References in the Proposal

Remember that throughout the proposal you must make a case for the importance of your research; in the introduction you can reinforce the relevance and need for the study. Use current and widely accepted references wherever possible to support your arguments and to channel the reader to your specific research objectives.

There is much uncertainty about the use of references in research proposals. Here are a few common questions (see also Chapter 12):

- How many references should I include?

- Which ones?

- Will references to hypotheses that compete with my work diminish my arguments?

- Should I use controversial papers?

• Do I need to justify my methods using the literature?

One of the first questions we hear concerns
how many references to include. The issue is the quality rather than the quantity of references. It is vital to illustrate or support your major points with important
references, but you do not need to include ten papers
illustrating each item that you raise. Include the references that have guided the development of the field,
and be sure to include any new references that are ger-
mane to your arguments.

The literature review must be considered thorough. Proposals are often criticized for not including
"key" references, which we take to mean the most
widely accepted or influential papers on a topic. And if
you are trying to establish that there is a gap in the literature in an area, you must be especially careful with
your literature review—you would not want to say
there were no published papers on a topic only to have
reviewers point out references that you missed. Another critical component of being thorough is the in-
corporation not only of references that support your
contentions but also of those that conflict with your arguments or are difficult to resolve with respect to those
arguments. New authors frequently ignore this aspect
of proposal writing. Yet many reviewers make a point

of considering both sides of all arguments, so this omission can be fatal to a proposal. Being thorough may also require that you cite controversial material. If you do so, be certain that you understand the reasons for its controversy in the field. Take special care if such papers provide the critical justification for your research (i.e., if you use them as a foundation for your own study).

Ask yourself if each concept or paper is worth citing. Bear in mind that you have limited space in which to develop your topic. Avoid unnecessary detail, parenthetical issues, and topics not specifically related to the proposal; they can be distracting, and they may suggest to reviewers that you have not clearly identified the key concepts and issues related to your proposal. We have even read introductions that endeavor to provide background on all aspects of a topic except the one the investigator plans to study! Imagine a proposal on the effects of climate change on soil microbial decomposition. If the author were to begin the background section by reviewing climate change theory, changes in greenhouse gases over time, and the evidence for and against global warming, it would be difficult to guess where the proposal was leading. There would not be an effective funnel, key arguments would not have been highlighted, and the reader could be well into the proposal before the relation between climate change, soil moisture, and soil microbial decomposition was established.

The Role of Models

The authors of some of the strongest proposals use conceptual, graphical, synthetic, theoretical, and analytical models to frame their research questions and design. Models, standard practice in some fields and less common in others, can be far more effective than words. Models are usually presented in the introduction and background sections and are often represented in the text with tables, figures, or sets of equations. It should be clear to the reader whether a model being used is your own or adapted from the literature. During the formulation of their research proposals, many scientists develop conceptual or analytical models that they publish as free-standing works.

Models have many different formats. Conceptual models are commonly used to identify the components of a study or the processes leading to and deriving from a central theme. A "box model"—a set of boxes and arrows that shows how your research question fits into a larger picture—may be effective.

Empirical models or calculations are also useful for synthesizing introductory and conceptual material. Many successful investigators apply a computer simulation or a few simple calculations to data from their own work or the literature in order to generate new figures, diagrams, or synthesized data that provide motivation and perspective for the current research. This is particularly effective when you do not have preliminary data but want to demonstrate that your ideas are feasible. The need to provide details about the model (e.g., specific equations, parameter sources, constants, hierarchical organization) depends on how fully the model predictions form the basis of your proposed research. However, you must ensure that you do not raise more questions than you answer by including the model. We urge you to ask colleagues who have used models to provide critical feedback on this topic.

Quantitative models can also be used to generate analytical solutions or predictions resulting from more formal mathematical expressions of conceptual or qualitative models. However, avoid models so complicated or untested that they are better suited to publish first in a peer-reviewed publication. The introduction is one potential location for a quantitative model; the methods section is another (Chapter 9). If the thrust of your proposal is modeling, then obviously you must present most of the details, probably in the methods section or in a section devoted to the model. If your model is a tool used to interpret or apply results, however, it is probably sufficient to present a short paragraph describing it, along with a few references to more detailed descriptions of how it is used. If you have modified a model from the literature but have not published your modifications, you should describe, if not show, your changes.

Are Objectives and Hypotheses Part of the Introduction?

It is common for the introductory section to conclude with a list and brief discussion of the objectives and hypotheses. These should derive easily from the background material preceding them. Some people prefer to include them in the research plan (Chapter 9) or as a separate element of the proposal; in either case, they must mesh logically with material in the introduction. Specific aims in NIH-type proposals do not need generally to be repeated. However, some authors organize the entire background into separate sections with headers for each aim.

In some instances, it is necessary to provide literature support for the employment of a particular approach or method, or for the use of a special type of instrumentation. If these are controversial or novel, the literature and preliminary results that endorse your project should be featured in the introduction and background. Your evidence for the successful use of these approaches will be of special interest to reviewers, who need to assess the feasibility as well as the importance of the proposed research.

Crafting the Introduction

We emphasize three issues when writing and organizing the introductory sections:

- Focus on the important points and establish their relevance to your proposal.

- Do not make this section too long.

- Use schematics, models, headings, and formatting to channel the reader and to show the direction the proposal is taking.

Introductory sections often are too long. A length of three to four single-spaced pages is common.

Remember, this would be nearly 25 percent of an NSF-type fifteen-page proposal. To take advantage of this limited space, you must choose your material wisely so as not to dilute your message. Organize your points strategically to channel the reader to your research, and break up text with headings, figures, and tables. Explain to the reader why you are presenting specific details by providing section and subsection headings, topic sentences, transitions, and written road maps such as, "In the following section we develop a framework for . . ."

Parts of the background may be nicely summarized with a figure or table. Figures are impressive if you have made them yourself and if they synthesize a number of different papers culled from the literature. Reviewers usually prefer to examine a figure or table embedded in the text rather than read two or three paragraphs of prose.

One final word of caution—resist the impulse to finalize the background too early. Novice writers may compose and finalize the background before they have finalized their hypotheses or specific aims. While it is great to gather essential background material early in the development stage, ultimately you will need to pare down and focus the background to your exact hypotheses and aims. It is often harder to rewrite or edit a section than it is to wait until you know what you want to include before writing it down.

Exercises for Writing the Introduction and Background

It is difficult to isolate exercises specific to writing the background or introduction of a research proposal. As usual, reading and critiquing other proposals and having others evaluate your own are the best preparation. It may be useful to consider good review papers as templates for an effective background. A strong synthetic review usually brings together the literature to support a set of central conjectures or to reveal a vital gap in knowledge. In this way it is similar to a proposal background.

The exercises that we presented in Chapter 4 lead to the development of the conceptual framework, the project summary, and underlying qualitative or quantitative models. These drills all require you to frame your questions within the context of the existing literature or of your own preliminary results, so they are also preparation for drafting this section. In particular, exercise 3 promotes the construction of a model or series of models to identify the key relations between

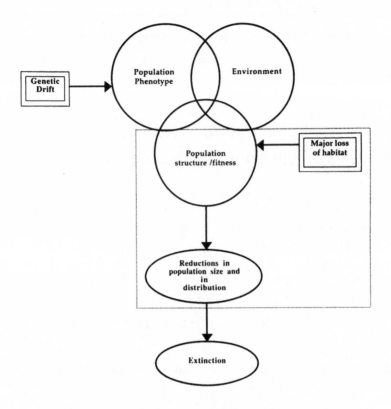

A model showing possible consequences of anthropogenic
habitat loss on population structure and fitness, distribution,
and extinction. In this example, the dotted lines indicate the
area that the hypothetical proposal will examine. Modified
from Gilpin and Soulé 1986.

processes that form the basis for a model in your introduction.

Finally, it is usually instructive to summarize
the elements of your research design schematically. Try
constructing several diagrams or figures that could
help you formulate a verbal description of your research plan. If this method proves effective, you may
choose to use this to help frame the introduction for
your project. If your introduction is a success, the
reader will be eager to reflect on the research plan that
you have devised to test your hypotheses and meet your
objectives.

Experimental Design and Methods: What Will You Actually Do?

The introduction and background should provide sufficient material to give the reader a solid appreciation of the importance of your proposed objectives, hypotheses, or aims. The next big step is to craft the unit referred to by NSF as the Research Plan. This element contains the nitty-gritty of the implementation, analysis, and interpretation of your study. This is where you convince the reader that your project can be accomplished.

Suggested Order of Presentation of the Research Plan

III. Project description (following from Chapter 3)
  A. Results from prior agency support (Chapter 8)
  B. Statement of the problem and significance (Chapter 4)
  C. Introduction and background (Chapter 8)
    • Relevant literature review
    • Preliminary data
    • Conceptual, empirical, or theoretical model
    • Justification of approach or novel methods

D. Research plan (this chapter)
- Overview of research design
- Objectives or specific aims, hypotheses, and methods
- Analysis and expected results (see also Chapter 10)
- Timetable (Chapter 11)

The research plan is usually broken down into a number of separate elements. In this chapter we cover the parts that deal with research strategy and methods. The trick is to keep from getting bogged down in detail. You must ask yourself, "Too much, too little, or just enough?" in writing this section of your proposal. There are several crucial questions that will be asked by thoughtful reviewers as they assess your design, methods, and analysis sections:

- Are these the correct and best methods for the specific questions?

- Are the methods proven and properly cited?

- Are the methods feasible given the time and support available?

- Is the precision or extent of the study appropriate and sufficient to answer questions, hypotheses, or objectives?

- Are the investigators competent in the use of all these techniques?

• What critical and innovative outputs will result
from this study?

Organizing the Research Plan

Organization of the research plan differs widely
among writers and among disciplines. The goal is to
keep the reader focused on the overall significance, objectives, specific aims, and hypotheses while providing
important methodological, technological, and analytical details. There are numerous ways to succeed at this.
Here we provide two examples that we have encoun-
tered in effective grant applications.

MODEL 1. In this model, the overall research
strategy and scope are presented in a short opening
section in a paragraph or two. Typical titles are Research Design and Scope, Experimental Design, Research Protocols, and Strategic Plan. This section
serves as a blueprint to keep the reviewer focused on
the research objectives. It can be particularly helpful
when organizing research plans comprising several
distinct sections that require different methods and approaches (e.g., Study System, Sampling Program, Analytical Techniques, Field Experiments, etc.). Describe,
at the start, your overall plan for linking these approaches.

Information must flow logically from the introduction and background sections. Provide evidence that your design is the best and most appropriate approach to solving the questions that you have identified as significant. Do not simply repeat what has already been said. Save space by including a table of key hypotheses, or a figure that conceptually links hypotheses and methods to specific objectives or aims.

Some authors follow with a section on the specific methods—Methods and Materials, Experimental Protocols, and so forth. The order of presentation of methods should parallel the order in which the objectives and hypotheses are presented elsewhere in the text. Tables for complicated procedures may be helpful, and appropriate referencing to common methods is essential. Be sure to make clear which methods pertain to which questions or hypotheses. A third section will probably be necessary to provide details on statistical analysis of the data.

Model 1 Format for the Research Plan, Methods, and Analysis

Research Plan
1. Overview of research plan and justification
2. Methods and materials
   • Sampling procedures
   • Culturing methods

   • Experimental protocol 1, 2, 3, etc.
   • Technical procedures, etc.
3. Data analysis

MODEL 2. In this model each objective and its associated hypotheses are presented with the study plan and methods that will be used to test them. For example, imagine that you have several objectives with their own derived hypotheses. In this format you present each objective separately, followed by the specific hypotheses and methods to examine them. This sequential approach can be effective in that each test follows directly from each question. However, in some cases you may find that you need to repeat material from previous methods, which may become tedious. With this format, authors may also include the analysis right after the methods section, eliminating the need for a separate section.

Model 2 Format for the Research Plan, Methods, and Analysis

Research Plan
1. Objective or Specific Aim 1
    Hypothesis 1A
        • Methods, materials, and protocol for H1A
        • Data analysis for H1A
    Hypothesis 1B
        • Methods, materials, and protocol for H1B
        • Data analysis for H1B

2. Objective or Specific Aim 2
   Hypothesis 2A
   • Methods, materials, and protocol for H2A
   • Data analysis for H2A
   Hypothesis 2B, etc.
   • Methods, materials, and protocol for H2B
   • Data analysis for H2B

What to Consider While Writing the Methods

Your research design and methods section is where you confirm for your readers that you have an achievable research plan. It is also where you highlight and defend innovative methods. Reviewers will carefully weigh this section in considering the pros and cons of the design. They will be asking questions like those that follow as they read your proposal.

Are these the correct and best methods for the specific questions? Perhaps the most decisive question that a reviewer will ask is whether the context and methods are the most fitting for meeting your objectives and testing your hypotheses. Although matching methods and objectives sounds simple, and it can be, some reviewers contend that inappropriate methodology is one of the most common flaws in unsuccessful proposals.

"I think I found your problem. You're putting in garbage."

Consider the following hypothetical example. An investigator proposed to measure the effects of aluminum on the reproduction and mortality of a particular species of fish. This author had convincingly argued the need for and value of this information, so reviewers were initially positive about the study. But as they read the design and methods section of the proposal they detected what became a fatal flaw for this study. The investigator planned to measure the total level of aluminum in fish tissue. Yet recent research had revealed that aluminum occurred in several chemical forms (species) in nature, and that not all forms were toxic to fish. Moreover, in some situations the toxic species of a metal has no relation to its total concentration. The reviewers concluded that the investigator would be making virtually meaningless measurements, and they now doubted the author's preparedness and understanding of current developments in the proposed field of study. In this simplistic example, the questions and hypotheses were sound, but poor methodology raised questions about the value of the work and the competence of the scientist, which resulted in a failed application.

Many other problems originate with a poor choice of methods, especially when there is no agreement in the field as to the best and most suitable approach. For example, the "best" (the fastest and most conclusive) procedures will generally be favored over others that also are feasible and workable, unless there is a serious drawback to their use. However, limited resources or access to particular instruments may prohibit your use of favored techniques. You may propose a less costly method for solving a problem, but it must be equally precise and reliable or reviewers are likely to object even if they are sympathetic to your plight. It would be beneficial to introduce the issues of cost of and access to new methods and technologies during the development of your ideas.

Are the methods proven and properly cited? Not all reviewers of your work will be experts in your field, so you must require evidence for the suitability of the proposed methods. Proper citation of established procedures and their application should suffice, eliminating the need to devote text to describing well-tried techniques. Innovative methodology and novel techniques can be highly regarded, but new or more controversial procedures require justification and affirming support. The most convincing support is documentation of their feasibility. We have submitted grant applications requiring adaptations or development of new techniques. These applications have been most effective when we could provide preliminary data or calculations to suggest the likely success of the procedure. In one case the proposal was awarded a small amount of funding (5 percent of the request) specifically to generate preliminary data to demonstrate feasibility.

The need for preliminary data is a well-appreciated conundrum in science—you need some results before you can be funded, but you do not have funds to conduct the work until you receive the grant. Collaborating with scientists already generating results of the type you desire is one solution, and writing small development grants is another.

Are the methods feasible given the time and support available? There are other ways that flawed methodology can be the downfall of a proposal. For example, in some cases the methods may be feasible—they will do what you say they will do—but impractical. What you propose to do must be reasonable given your time and resources.

Some people seem to feel that it is better to err on the side of proposing too much rather than too little. Graduate students often conceive projects that would require several dissertations to complete. While this may be acceptable to a tolerant graduate committee, a funding agency is unlikely to accept a serious mismatch. Proposing too much casts doubt on your judgment, and it does not provide reviewers with information on how you prioritize the tasks. It is more effective to educate the reviewers and convince them that you can do what you say you will do. The timeline for a proposal (Chapter 11) may be scrutinized by reviewers for this purpose.

Is the precision or extent of the study appropriate and sufficient to answer questions, hypotheses, or objectives? When you have established that your questions will be tested properly by your proposed methods, critical reviewers will assess the way you intend to handle your data and conduct statistical analyses (see also Chapter 10). An effective proposal demonstrates that you understand and propose to use the best and most powerful analytical techniques suited to your specific experimental design, sample size, and replication. It will be useful to consult with colleagues and advisers about the data analysis. In some disciplines, mistakes in data analysis are among the top reasons that a proposal fails to be funded.

Are the investigators competent in the use of all these techniques? Reviewers will query whether you and your collaborators have the technical expertise to accomplish the research. In most areas of study there are laboratory techniques or field methods that are difficult, expensive, or time consuming. If you intend to use these means, you must prove not only that they are appropriate and feasible but that you are qualified to use them. The testimony of people with whom you have studied or trained, your preliminary data, and your peer-reviewed publications will be used to evaluate your competence with a given technique or method. If you are trying something novel and are also new to a discipline, you may be unable to prove your skill in a proposal. You may wish to collaborate with someone who is well regarded in the field (Reif-Lehrer 2005). It may be enough to demonstrate that the "expert" has agreed to advise you or to conduct calibration and cross-checking of samples, methodology, or techniques. A more formal collaboration (such as a subcontract) may be best in circumstances where your ability to conduct difficult work may be questioned. Many agencies encourage multi-disciplinary collaboration, and this is an excellent justification for working with someone who complements your abilities.

What critical and innovative outputs will result from this study? Every effective proposal summarizes expected outputs. It is common to see proposals with a special section titled Expected Results (Chapter 10). The projected outputs of your study must be considered realistic and important. In some cases, investigators include a specific set of items that they will pro-

duce (a gene library, a collection of specimens, and new
drug therapy), while others generate less tangible outputs (advancing our knowledge of a topic, testing an
untested theory). Outputs should follow directly from
the significance statements and in some sense address
how the actual study is pertinent to the broadest stated
goals. Be aware that reviewers may assess your work
with this in mind.

Deciding What to Include

The methods and experimental design section
of a research proposal differs from that of a manuscript
in a fundamental way: proposals must include enough
information for a reviewer to critically evaluate your
methods, but they do not have to provide all the detail
needed for someone to duplicate your work. There are
times when you should list specific methods—particularly if you are relying on unpublished or novel procedures—but often it is appropriate simply to cite an earlier work or standard technique. Presenting too much
information can diminish the punch of your proposal,
but presenting too little may prompt readers to question your ability to conduct the work. Consultation with
colleagues or even the program director at specific

agencies can help you decide what is appropriate in your case. Just as you evaluate the contribution of each reference (Chapters 8 and 12), so should you question the need for each detail of your methods.

All proposals require a distinctive set of measurements or experiments, so it is not possible to produce a checklist of items for this portion of a research proposal. A few elements common to most methods sections are listed below; refer to proposals in your own area for more specific information.

Outline of proposed research. Most proposals include an outline or brief description of the overall methodological approach. This functions both as a road map and as a justification of your approach.

The study site, the species, the system. If your work depends on a certain organism, location, gene, model system, chemical product or process, or on a specialized piece of equipment, you should describe this here. Supply adequate information to allow the reader to understand the system in which your work will take place. This is especially important for reviewers who are not experts in your field. Discuss the background of the reviewers with the program director to determine how much detail to provide.

Methods. If your techniques are well known, it may suffice to write, "Collection and analyses will be

conducted following the methods of So and So (1997)."
If the methods are new, have been developed by you, or
are not well known, thorough description and documentation are essential. Do the proposed methods
have any particular limitations that might affect the
interpretation of your results? Explicitly state those
limitations and their implications. It is much better
for you to point them out than for a reviewer to raise
them.

Data analysis. If preliminary data are available,
show how you will use or analyze them to reach your
objectives or test your hypotheses. If such data are un-
available, consider culling data from the literature to
show how you expect the results to turn out, and to
show how you will analyze your data when they are
available. Complete a table or diagram, or run some
statistical tests using the preliminary or "synthesized"
data. This can be a good way to show how you would
interpret the results of such data (see also Chapter 10).

*Sample/data storage and archiving; integrity of
data.* If appropriate, describe your intention to save,
store, or archive physical samples or data (and any specific methods that you would employ). There are a
number of situations in which this is beneficial. For example, you may be required by your funding agency to
make data or samples available to other investigators.

Even if these are not required, making your data accessible to other investigators may strengthen the proposal. You may also be planning a follow-up or comparison study. If so, maintaining the integrity of the samples or archived data base is critical.

Exercises for Writing the Methods Section

Critique other proposals. Once again, our first suggestion is that you critique other proposals in your immediate research area. Consider the extent to which they provide information and the way that it is presented. Act as a reviewer, using the questions listed above. Examine both successful and unsuccessful proposals, if possible. Unsuccessful proposals may provide some guidance on ineffective presentations or perceived weaknesses. Some authors may be willing to share samples of reviewers' comments from successful and unsuccessful proposals. Often the reviewers are explicit about the key weaknesses, which frequently occur in the methods and approach sections.

Critique these excerpts from the research design sections of two funded proposals.

Once the precise bases that interact with the gene APETALA3 (AP3) are identified, we plan to make specific base changes by site-specific mutagenesis (Kunkel et al. 1987). Once the mutations are made, we will clone them back into an otherwise wild-type AP3 promoter fused to b-glucuronidase (GUS). These constructs will then be transformed into the plant *Arabidopsis thaliana* and crossed to see whether they activate the GUS reporter. If single base changes destroy the ability of AP3 to autoregulate, that would provide convincing evidence that the mutated sequence mediates autoregulation.

Source: T. Jack

Within each section of stream we will survey three to four channel cross-sections (Feldman 1981) from valley side to valley side, for subsequent hydraulic modeling. Surveying will be accomplished with a Topcon AT-F6 automatic level. We will measure gravel sizes on riffles, channel area, pool area, and bar area using standard techniques and those of Wolman (1954) and Hankin and Reeves (1988). . . . The effects of channel bed roughness (primarily a function of bed particle size and slope) will be determined using empirical relationships determined by Jarrett (1985) for steep cross-sections typical of high-elevation streams.

Source: P. F. McDowell and F. J. Magilligan

EXERCISE 6. (This follows from the exercises given in Chapters 4 and 7.) Prepare a ten- to fifteen-minute oral presentation on the research design and methods for your research project. As in exercises 1–5, emphasize the relation of the study to the larger field of theoretical and empirical research preceding it. It will be difficult to do this in fifteen minutes, but by attempting it you will be forced to be succinct and precise. You may wish to undertake this drill several times, as logical flaws or poor presentations are revealed by feedback from colleagues or classmates. Soliciting comments from people in more peripheral disciplines can be especially helpful for distilling the jargon from your presentation.

After you work on your own methods section and research plan, seek comments on your written work. To help those giving you feedback, you may even suggest a list of questions (e.g., Is my description of how I will sample clear? Do you understand what I plan to do with the data?). After completing this section to your satisfaction, you have nearly completed your proposal. Your remaining tasks are much less momentous.

Plan for Expected and Unexpected Results

Strong scientific research proposals usually include a section following or embedded within the research plan where the author presents the expected results and explicitly discusses their interpretation. However, even with the most carefully designed proposals one can face obstacles during implementation, or produce unexpected results that require rethinking original precepts, redesigning experiments, adding new or eliminating parts of original protocols.

In our surveys of colleagues, we were struck by the emphasis they placed on the importance of investigators considering both likely and unlikely outcomes. They felt that investigators who are prepared to rapidly redirect research, or quickly respond to unusual yet important results, often produce the most exciting results. Science is filled with dramatic examples of a major breakthrough coming serendipitously through a failed

"WE'RE AT THAT AWKWARD STAGE
OF TRYING TO EXPLAIN A COMPLETELY
ACCIDENTAL DISCOVERY AS OUR
INTENDED OBJECTIVE."

experiment or as a byproduct of an unrelated project. Therefore, we recommend that you also discuss unexpected or unlikely outcomes along with expected results, and we offer a few ways to address these issues in your proposal.

Explicitly Address Outcomes

Careful consideration of expected outcomes underlies scientific competency and shows that the proposal authors are well prepared. Often the outcome of a project or experiment is fairly predictable. This does not mean that the research is stale or boring but may simply indicate that the authors effectively established hypotheses and models, examined the literature for applicable results from other systems, and perhaps were just a bit lucky.

There are many ways to explicitly address outcomes in your proposal. Try several approaches.

- Graphically depict the relationships you expect and discuss your analysis and interpretation. Be sure to briefly discuss the interpretation that you will make if the patterns do not come out as predicted. State what such unexpected outcomes mean for your overall goal. In many cases, writers include fairly detailed descriptions of the analysis necessary to interpret the results. Attention to analysis can be especially important when the analysis is novel, difficult, quirky, or somewhat controversial.

• Provide diagrams of the approaches you plan to use for different outcomes. If everything you propose depends on one specific outcome, and that outcome does not occur, then the research will be considered very risky. Make clear that each pathway leads to valuable and interesting results. Although risky research is sometimes necessary (i.e., when the gains are worth the risk), when alternate outcomes all lead down interesting pathways the research may be more likely to succeed.

• Construct a simulation model to predict likely outcomes (see also Chapter 8).

• Estimate the likelihood that alternative outcomes can result. Some proposals include a table with alternative hypotheses or outcomes. In certain cases it is possible that the alternative requires a mechanism that defies all logic. If so, little explanation may be needed, compared to a case where two mechanisms are nearly equally plausible. It is worth anticipating unlikely events, as reviewers from different perspectives or disciplines may consider issues that you do not. By considering alternatives you may be able to identify aspects of your research that require special explanation for a broad audience.

Given other constraints on your proposal, your discussion of expected and unexpected results will probably be restricted to a few paragraphs or figures. The extent to which you need to discuss this depends in part on the ramifications of the different outcomes to your research design, hypotheses, and broader goals.

There is no consensus from authors and reviewers as to where this material should be located within the proposal. Some writers discuss results throughout the methods and hypotheses sections, while others place the discussion near the end of the body of the proposal. Sample titles for such sections include:

- Expected Results and Their Broader Significance

- Future Directions

- Related Research

- Model Limitation and Potentials

- Model Verifications

The Timeline Is a Reality Check

An organized and simple timeline is useful to both author and reviewer. Devising a timeline helps you acquire an appreciation for the links between tasks and for the time required for each part of the study, and insight into the resources (money and labor) required to complete the proposed project. This thinking process makes it more likely that you will budget correctly to successfully carry out the project. At the completion of a grant cycle you may decide to reapply for funds to continue your work. If you have underestimated the time and budget needed for the original project you will be in the position of justifying yourself in the new application.

Although the formal construction of the timeline is often left until after most other sections of a proposal have been written, it is best to think seriously about the time required for specific tasks while developing your research plan. A well-conceived timeline demonstrates to reviewers that you have carefully assessed the personal and financial commitments entailed by your project. A realistic timeline shows that the project is feasible and builds confidence in your judgment. If reviewers find the timeline overly optimistic they may have doubts about funding the proposal. Scientists with experience conducting the same type of research as you propose are your best sources of assistance in this area.

Constructing a Timeline

In NSF-formatted grants, the timeline usually appears at the end of the project description (see Chapter 9), just before the references. There are no blueprints for a timeline, but the start and finish of milestones in the project are usually depicted in chronological order. Sometimes authors simply list the general tasks and target dates, while others use more elaborate schematics. In general, the timeline is no more than one page long. If requirements for particular tasks need justification, include it in an accompanying paragraph.

Many different types of information are included, such as:

- the beginning and end of each field season or experimental period;

- time needed to construct an instrument, purchase new equipment, or develop a novel technique;

- time needed to create a genetic library, grow particular cultures, etc.;

- time scheduled to use equipment at other facilities (e.g., telescope time, use of a mass spectrophotometer, sample deep-sea squid, etc.);

- starting and completion dates for a monitoring program;

- when to expect publication of results.

No matter how carefully you plan your timeline, at some point "science and reality will meet," as a friend of ours says. All of us, even veteran researchers, can underestimate the time required for different tasks. A good guideline to follow is to make your best estimate of how long something will take and then double it. This still may not be enough time, but it may put you in the ballpark.

Here are two examples to consider before drafting your own timeline.

Sample 1

Timetable for a Fictional One-Year Study on Amphibian Density and Respiration

JUNE 2009

- Order equipment.

- Receive, test, and calibrate equipment.

- Choose sites and test sampling methods.

- Install temperature monitoring stations.

JULY–OCTOBER 2009

- Collect continuous temperature data.

- Make weekly amphibian collections and measurements.

- Make monthly assessments of amphibian densities.

NOVEMBER 2009–MAY 2010

- Analyze data, perform statistical tests.

- Construct respiration/density model.

- Prepare papers for publication.

- Submit final report.

Sample 2

Timetable for a Four-Year Study on Extracellular and Intracellular Signals Regulation Cell Division and Differentiation

Projected Timetable	Year 1	Year 2	Year 3	Year 4
Cloning of *pex-1*	xxx	—	—	—
Molecular investigations of *pex-1* function	—	xxx	xxx	xxx
Genetic characterization of *pex-1*	xx	xx	xx	xx
Studies of spatial regulation of RAS-MAPK activity	—	xx	xx	xx
Screening for pachytene exit mutants	xxx	xxx	xx	x
Screening for suppressors / enhancers of RAS-MAPK	x	xx	xxx	xxx
Characterization of newly identified *pex* genes	x	xx	xx	xxx
Dissection of cellular regulation of pachytene exit	xxx	x	—	—
Examination of PKA and cdc2 roles in pachytene exit	xx	xxx	xx	xx
RAS-MAPK and mitogenic signaling	xx	xx	xx	—
Investigation of the UCN fate	x	xx	x	—

— = done or not begun; x = low percentage effort; xx = moderate percentage effort; xxx = high percentage effort
Source: Lambie, Dartmouth College

References in Detail: How Many and How Recent?

Properly cited and appropriately chosen references are essential to a strong proposal. References alone will not determine whether a proposal is funded, but weak use of citations suggests insufficient preparation and undermines confidence in an otherwise good application. Citations convey critical information to the reader with a minimum of words. Inexperienced writers sometimes have difficulty knowing how often and where to position references, and how to select among many possible choices. Below are guidelines for the judicious use of citations in a proposal and a review of some of the more common referencing problems.

The Basics

Which references to cite? This is a fundamental question for authors of research proposals and manuscripts. The selections you make indicate a great deal about your perspective and knowledge of the current state in your field (see also Chapter 8). It is advisable to cite papers that are well known for their importance to a specific topic. References addressing your system directly and perhaps even your specific research question should also be used. We again emphasize the importance of being unbiased: cite papers that dispute your position and directly discuss the key differences in opinion. You should probably rely most heavily on recent papers, although in certain cases an older paper may still be considered the seminal reference. In general, most cited papers will be less than ten years old, and inclusion of extremely recent references lends freshness to your application (and demonstrates that you keep current with the literature).

The conventions for citations in manuscripts also hold for proposals. For example, cite peer-reviewed works whenever possible. Use nonreviewed work, reports, unpublished data, and personal communication sparingly. Cite your own work but not excessively.

Reviewers always weigh whether your proposal contains the most appropriate and consequential references. Be sure that the answer is yes (see Chapter 8). Consider who is likely to review your proposal (see

Chapter 1). Remember that the scientists who will evaluate your study are probably doing some of the best
and most relevant work, so it is reasonable to include
one or more of their papers.

Finally, cite only papers that you have actually
read. Occasionally writers lift lists of references from
the text of other papers. This can be dangerous, as often the citations will not be relevant in even a slightly
altered context. Reviewers do pick up on such careless-
ness.

How many references should I include? The typical answer—"You don't want more than you need, but
be sure to have enough"—is vague. In general, cite all
papers that are essential to establishing credibility or
feasibility, but limit the use of citations that simply provide background and support. When establishing background, include the few most important or influential
papers. Lists of more than five papers following a statement are rarely necessary. Do not use more references
if one will suffice. Numerous citations for facts that are
well appreciated in your discipline suggest naïveté on
the part of the writer.

All citations should be accurate. Your references
must be correct—pay particular attention to the date of
publication. One of us was chastised for having a few
references in the text that did not match the full cita-

tion at the end of the paper (i.e., the year of publication was not identical in both places). A reviewer wrote that since he or she was unable to see the data generated by the research, references were the only information that could be checked to determine whether the authors were likely to be careful and precise. The reviewer said that if a mistake in the citations was representative of the care we gave to our work, our research may not be scrupulously done. A bit harsh (we thought!), but indicative of scientists' strong feelings regarding perceived sloppiness.

Common Problems

Here are some simple examples that illustrate how placement and number of references can lead to ambiguities. All references are fictional.

Too Many References, or References Are Too Vague

"Air pollution affects plants in a variety of ways (Browner and Bowles 1994, Kramer and Berger 2001, Smith and White 2003, Pearce and Omer 1999, McPhee et al. 2006)." This is a vague statement to

begin with, which makes it difficult to reference. It is
not clear what the references in this list are supposed
to indicate. If you use such open-ended statements, select only one or two general reviews for support.

"Air pollution affects plants in a variety of ways
(e.g., Smith and White 2003)." This suggests that
Smith and White is a review paper that discusses the
many ways in which air pollution affects plants. If
Smith and White present only one way in which air
pollution affects plants, it is inappropriate to reference
the statement this way.

References Placed Poorly in the Text

Some newer authors tend to place all references
at the end of a paragraph when references may refer to
different clauses within the paragraph. This can be
very confusing and dilute the impact of the reference.
Note:

> While studies of the effects of vines on riparian
> plants have been conducted, there has not be an in-
> tegrated analysis of the effect of vines on riparian
> plants (Asanti and Laszlo 1991, Schwartz et al.
> 2004, Jones and Smith 2005, Fullington 1977).

versus

> While studies of the effects of vines on plants
> in deltas (Asanti and Laszlo 1991, Fullington 1977)

and estuaries (Schwartz et al. 2004, Jones and Smith 2005) have been conducted, there has not been an integrated analysis of the effect of vines on riparian plants.

The first sentence suggests that four studies identified a need for an integrated analysis of the effects of vines on riparian vegetation. In the second sentence, specific studies in deltas and estuaries are cited. The absence of a reference at the end of the sentence indicates that no one prior to the author identified the lack of an integrated analysis of vines on riparian plants.

Weak or Incorrect References

By including a paper in your references you are saying that you have read the paper and understood it thoroughly. Cite it carefully, as authors are usually perturbed when their work is referenced incorrectly.

Ambiguous Use of Reference Notation

Do not confuse references that are examples of your points with ones that support your argument in some other way, or make a similar point to the one you are making (e.g. = such as; i.e. = that is; cf. = confer).

Exercises for Writing References

There are several activities for preparing to choose references:

- Check the citation format for the agency to which you are applying.

- Develop a library of references for your proposal using packaged software.

Here is a simple drill. Read a paragraph written by a colleague or classmate in a field that is unfamiliar to you. Examine each reference and speculate on why it is there and what it should be telling you. Check with the author to see whether the intended message was correctly interpreted. This exercise is most effective when you are not familiar with the references cited. Apply the same consideration to your own citations.

Preparing a Budget

Most scientists write their budgets after completing the rest of their proposal. Whether you are applying to NSF for $250,000 or to your department graduate student fund for $250, there are some basic principles to follow when preparing your budget. The most important of these is to consider the ethics involved in accepting financial support for scientific research. You are obliged to follow the terms of the award, to include costs that relate directly to the research, and to take full responsibility for the veracity of your data and the appropriate use of the research dollars.

Budget writing requires careful planning and detailed knowledge of institutional administrative and overhead costs, and fairly precise estimates of equipment purchases, research assistant time, costs of supplies, and anticipated travel. For ethical and practical reasons, we suggest that you provide your most accurate estimate of the cost of conducting the research. Although some agencies are flexible in the extent to which they allow award costs to be transferred among categories as the research progresses, others are much stricter. In either case, the closer you follow your original budget, the better.

Virtually all federal agencies require that budgets for proposals be submitted through www.grants .gov, so the budget will almost certainly be submitted by your grants department rather than by you.

What Do I Include?

Preparing a budget is a process that gets easier with experience. All federal agencies, and most funding organizations, have a page for formally itemizing the budget into categories. These organizations are usually willing to provide detailed guidelines for particular costs. Discuss these issues with both your own campus grant office and the program director at the funding agency. Do this well in advance of the target date for submission, because getting a budget finalized and ready for submission to the funding agency can be complicated. Most agencies require a page bearing your signature and the signature of the individual at your institution responsible for the oversight of the funds. You may have to obtain the signatures of chairs, deans, and provosts before the institutional representative will sign your budget page.

In addition to the direct costs of conducting research, the so-called indirect costs, or overhead costs, must be included. Each institution negotiates these rates individually with most funding agencies, and so your campus will supply you with the specific indirect costs associated with different budget categories (e.g., salary, travel, supplies, etc.). Occasionally agencies set rates that differ from your institution's, and you will need to discuss this with your campus grant officer. In addition, most campuses charge different overhead rates for different types of research (e.g., off-campus versus on-campus activities). Last, most agencies restrict the overhead allowed on large equipment purchases. For all this information, contact your institution's grants office.

It is often necessary to include an additional page in which you justify in a few paragraphs certain budgetary requests. This budget justification page might explain why you need four thousand test tubes or sixteen trips to your field site during one field season. Here are a few of the most common budget categories.

"I'M AT THAT AWKWARD STAGE IN THE PROPOSAL WHERE I HAVE TO ASK FOR SOMETHING."

- *Salaries.* People are expensive, and research is labor intensive. For principal investigators, one-half, one, or two months of summer salary is typical; some programs also have salary caps, and some programs do not allow PI salaries. For other research staff, some number of calendar months should be indicated, depending on the percentage of a person's time that will be devoted to the project. Graduate stipends and undergraduate salaries are also often included. Each salary category carries different associated costs to cover benefits and overhead. There are many regulations—some old and some new— regarding salaries and benefits, so always rely on your institutional grant management people to assist you with this.

- *Equipment.* Most institutions and agencies have specific definitions of equipment (e.g., those items that cost more than $2,500 and will last more than one year). It is appropriate to include in your proposal equipment that is necessary to conduct the proposed work. However, if it is prohibitively expensive, it may limit your success. Any significant equipment expenditures should be discussed in advance with the program officer at the funding agency to determine their suitability. For large purchases, institutions and agencies will sometimes "cost-share" agreements, with both parties combining funds to purchase equipment. But plan in advance, because these arrangements must be finalized prior to submission, and they require time to negotiate.

- *Supplies.* Include all expendable items, including lab ware, chemicals and reagents, field utensils, and minor computer accessories necessary for the successful completion of your project.

- *Travel.* This category should include the destination and number of trips, the mileage, the cost for lodging, and a per diem for food. Justify all travel destinations (e.g., field sites, meetings with collaborators, conferences, outreach, patient reviews, etc.). Obtain your institutional guidelines for costs related to travel. It is common to include travel costs for presenting results of the study in later years of a project, but again each agency will have its own rules and restrictions.

- *Miscellaneous expenses.* These are not part of the institutional costs to be covered by the indirect cost fee. These expenses may include computer time or courier services for rapid delivery of time- or temperature-sensitive samples. Discuss these items with other people in your department and grants office to determine rules and common practices.

- *Subcontracts.* This category applies when you plan to pay for work conducted at another institution. Again, local regulations will apply.

Cost Sharing

It is increasingly common for funding agencies to ask for matching funds or cost sharing from the investigator's institution. Some common examples of cost sharing include graduate student stipends, equipment purchase, or undergraduate internships. Cost sharing can often demonstrate institutional support for a particular proposal, and it can benefit others at the institution beyond those directly involved in the proposed project.

Final Thoughts on Preparing Budgets

Most scientists carefully estimate the true costs they expect to encounter while conducting research, and they submit a budget based on those realistic estimates. We strongly advise adopting this strategy. However, it is easy for those early in their careers to underestimate the amount of work involved and, consequently, to underestimate the amount of money needed. Caution, care, and consultation with senior colleagues or advisers are important. Preparing a fair budget not only will make your proposal more competitive, but it is the ethically appropriate strategy.

	Itemization	
Sample Budget for Field and Lab Study	Year 1	Year 2
Principal investigator,		
1 month summer salary	$8,431	$8,853
Fringe benefits @ 30%	2529	2,656
Postdoctoral research associate,		
full time	45,000	47,250
Fringe benefits @ 25%	11,250	11,813
Graduate research assistant,		
full time	20,000	21,000
Fringe benefits @ 15%	3,000	3,150
Part-time laboratory technician,		
12 hours per week	7,488	7,862
Fringe benefits @ 10%	748	786

Sample Budget for Field and Lab Study	Itemization	
	Year 1	Year 2
Summer field and lab assistant,		
40 hours per week for ten weeks	4,000	4,000
Fringe benefits @ 10%	400	400
Travel total	1,650	1,650
Ten trips to field site (100 miles		
round trip $.45/mile)	450	450
Attend one conference	1,200	1,200
Supplies total	2,000	2,000
Reagents	300	300
Field sampling bottles, bags, and tools	500	500
Laboratory chemicals and containers	700	700
Analytical chemicals	500	500
Publication costs	1,000	1,000
Equipment	0	0
Total direct costs	107,496	112,420
Indirect cost @ 50%	53,748	56,210
Total budget	161,244	168,630
Total request for years 1 and 2		$329,874

Submitting and Tracking Your Proposal

If you have finished your proposal, congratulations! You now have checked it for errors, typos, and formatting, and made certain that you followed the guidelines and rules of your institution and the funding agency. You've made the required number of copies or examined the guidelines for electronic submission. While some private foundations and state agencies accept hard copies of proposals, only electronic submissions are accepted at most federal offices, including NSF and NIH. For the foreseeable future, investigators who receive awards from NSF will continue to use FastLane (the U.S. government's electronic submission gateway) for ancillary activities such as submitting annual reports, revising budgets, and submitting supplements. But the federal Web-based portal www.grants.gov is the primary gateway between an institution and the federal government.

As you were working on your proposal, you presumably were also determining the proper submission method—paper or electronic—for the agency you targeted. For paper submissions, you must determine important issues such as numbers of copies, recommended fonts, and dates for receipt or postmarking. The trend is toward electronic submissions, however—even for certain foundation and graduate student proposals—and our most important piece of advice is to start well ahead of the submission date, particularly if this is your first time using a particular electronic gateway. Collaboration with multiple investigators is actually much easier with a networked system such as FastLane, which allows one investigator to upload a draft of a document and collaborators to access and modify it. Grants.gov may have this capability soon. Other tips about electronic submissions are best learned from colleagues and the grants or sponsored-projects department of your institution.

Although not all agencies require a cover letter, we suggest that you consider writing one. In a hard copy submission you are free to include a cover letter. FastLane and grants.gov allow submission of additional material, and, depending on the circumstances, a cover letter could be useful. Here you may describe the importance of the work, help categorize it if it is in a hard-to-identify area, or discuss any dealings you may have had with the program director. You may also list individuals who would be suitable reviewers of your proposal. (None of these people should be collaborators, advisers, or advisees.) In rare cases it may be reasonable to include the name of a reviewer to avoid if you have a specific reason. This information should be held confidential by the agency, but use the option sparingly. Most scientists never need to do this.

If submitting by mail, we strongly recommend that you use a service that will allow you to verify delivery. It is useful to know that your package arrived on time. It would be extremely unfortunate to go through all the work involved in preparing a proposal only to learn that it arrived after a deadline. With electronic submissions, your granting administration agency, which will submit the actual electronic document, will receive acknowledgment that the document has been successfully transmitted.

What Happens to Your Proposal After You Submit It?

Until the advent of electronic submissions, all proposals were received in the mailroom of the fund-

ing agency. Tens, hundreds, or even thousands of submissions may have been received within a short time
period. Regardless of how a proposal is received today
(through the mailroom or via an electronic submission), each proposal is assigned a tracking number or a
tracking link. If www.grants.gov is used as the submission gateway, the system immediately generates a
tracking number for the proposal. The tracking number is also provided in the application submission con-
firmation email, which is sent to the "authorized organizational representative" (usually someone in your
grants office). After validation by grants.gov, the proposal is forwarded to the specific federal agency, where
it will be assigned an agency-specific tracking number.
If your proposal is successful, this number will become
the award number.

After a proposal is processed it will usually be
sent to one or more individuals for review. Some agencies or private foundations conduct in-house reviews. It
is more common for the person responsible for evaluating grant proposals and making awards to assemble
a list of ad hoc reviewers who will provide a written review of your proposal. Ad hoc reviewers have no formal
association with the funding agency or its panels; they
are members of the research community who agree to

review a proposal. They perform a similar function to the ad hoc reviewers of manuscripts submitted to journals. Ad hoc reviewers are chosen many ways (e.g., from your citation section, your cover letter, past contact with the agency, general standing in the field).

Many agencies arrange for the ad hoc reviews to be evaluated by a panel consisting of scientists (often leaders in their fields, some of whom have received funds from that agency in the past) and members of the agency staff. For some U.S. federal agencies, several panel members read each proposal and write additional reviews that are added to the reviews submitted by ad hoc reviewers. A single panel member may be identified as the primary reader to present the major ideas of your proposal to the rest of the panel and comment on issues raised in the ad hoc reviews.

The panel discusses each proposal and generally writes a panel summary, which is a synopsis of the positive and negative aspects of the work. Panel members may also comment on the qualifications and productivity of the investigators. Proposals may then be ranked or placed in categories (e.g., "must fund," "possibly fund," "don't fund"). The program director generally makes the final decisions on funding and notifies investigators about the outcome of their proposals. Most agencies say when decisions will be made, and it is rarely a good idea to call or email the agency prior to this date. Many agencies do not make final decisions until four, five, or even six months after the deadline for submission.

People receiving good news usually receive it first, perhaps within a week or two of the decision. Bad news comes more slowly. Eventually you should receive a set of reviews of your proposal and a written panel summary. But we urge you to call the agency for feedback on failed proposals rather than wait for this packet, which can take months to arrive. We suggest calling the agency any time after you've received the decision, digested the fact that your proposal won't be funded, and are able to talk about it in a receptive manner.

If your proposal was funded, congratulations! If not, *don't despair.* In this era of limited funding and highly competitive programs, many proposals are not funded on their first submission. You will probably wish to revise and resubmit your proposal to the same agency. Careful consideration of the panel summary and reviewer comments is an important part of the resubmission process.

CHAPTER 15

The Three R's: Rethink, Revise, and Resubmit

Do not be too disappointed if your proposal is rejected the first time it is submitted. This is not uncommon, and a revised proposal is usually much stronger than the first. Some granting agencies provide you with several ad hoc reviews and a written summary of the panel deliberations. Consider yourself lucky if you receive a large number of reviews; it is a great advantage to receive plentiful feedback on your research. All scientists benefit from reviewers' comments and insightful critiques. Successful grant writers use these comments to revise and improve their plan.

Research proposals are rejected for many reasons, and even some very good proposals go unfunded. To better your chances the second time around, consider seriously each suggestion from the program director, agency head, or review panel. Remain open-minded—you do not need to agree with or adopt all of their suggestions, but at least be prepared to support your viewpoint in the next version. Below are general strategies for rethinking and revising your proposal. Your close colleagues or advisers can advise you about responses to more specific comments.

Significantly Improving the Resubmission

Many of us spend hundreds of hours on our proposals, so it is difficult to receive a rejection, particularly when the grant you have labored long to produce is brushed off in a few paragraphs. It is not unusual to feel that a reviewer or even the entire panel misunderstood something you wrote. Experiences like these can be useful if you take the position that they did not understand because you did not make your point clearly. Let your disappointment, anger, and frustration subside before you rethink and revise your proposal. By being as objective as you can you will improve the chances of getting your proposal funded upon resubmission.

Evaluate reviewers' comments. Begin by reading your proposal from start to finish. It has probably been a few months since you last read it, so you will see it

with fresh eyes. Look again at the reviewer comments.
As you examine each remark, try putting it into one of
the following categories: "must consider," "may consider," "ignore."

If you feel that all of the comments belong in
the third category, you are probably not accepting criticism very well. But if the concerns are distributed
among all three, or are primarily in the first two categories, you are off to a good start. Review the comments from the first two categories and identify the
sections of the proposal to which they are most pertinent. Observe the patterns that arise. For example, you
may notice that a large number of issues deal with your
methods or hypotheses. Or you may find that most
concerns relate to the scope of your project. This exercise will help give you a feeling for which sections of
your proposal need the most serious revision or were
least clearly expressed.

Edit your proposal. The extent to which you need
to rethink, revise, and rewrite differs for every proposal. Our rule of thumb is to adopt a strategy similar
to that used to conceive and draft your original proposal.

• Start with the big questions: Were they clear?
  Did the reviewer accept their significance?

Were the links between your broadest goals and your proposed research plan convincing? If these areas received the greatest criticism you'll need to rethink your proposal from the fundamental tenets.

• Funnel downward to the specifics of your research plan. Determine whether your discussion led to misunderstandings in the methods. Did you overlook important papers that could alter your research design? Did you make any mistakes in analysis or interpretation? Did you omit a key set of experiments that could change your final product?

• Look for strategic errors that could have affected the reviews. Were you unrealistic in your timeline? Did you fail to highlight key points, or did you overwhelm them so that they were lost? Was your budget out of line?

• Discuss the criticisms and your responses with colleagues.

Finally, remember that reviewers sometimes make suggestions that are not tenable or are off the mark, but if you follow this process you should be able to substantially improve your proposal. You have an

important added advantage over your first submission—you now know what some of the reviewers' concerns are likely to be. Make use of this information in
your revision. It can only help.

Writing a Resubmission Response

Most agencies require a resubmission response
as part of proposals that have been rejected previously.
Some agencies even require an item-by-item response
to prior reviews. A resubmission response should be
no more than a few paragraphs, since it is usually
counted as part of the main body of the proposal and
falls within page restrictions. As with responses to
peer-reviewed publication submissions, it is advisable
to summarize the major comments made by the reviewers (positive as well as negative) and describe how
these comments were incorporated into the revision or
why you chose not to assimilate them.

A resubmission response is valuable for a number of reasons. First, it allows the author to carefully
consider each reviewer comment and to check your response to it. Second, it guarantees some institutional
memory for your proposal. If the initial reviewers

agreed that you wrote an excellent proposal and that the only major weaknesses were in the methods, you should remind the panel of this in your revised submission. This response puts some pressure on the agency to acknowledge the previous review, even if there are new program officials and reviewers of the revised proposal. Third, it draws attention to the improvements you have made.

Some agencies require placing the resubmission response at the beginning of the project description, and others require placing it just before the references. Depending on the agency, you may also place it in your cover letter. When written well, it can be used to highlight key points or deflect key sources of potential concern, so placing it early may be advantageous.

Consider Private Foundations for Funding of Innovative Research

A question we often hear from colleagues is, Where can we find support for innovative work that falls outside the purview of more traditional federal research agencies? Increasingly, the answer can be found by turning to private foundations. Foundations in the United States already are funding a significant amount of research. For example, it is estimated that in 2005 more than $5 billion was awarded by private foundations to investigators in the sciences, environment, technology, and health care areas. Each year, more and more personal wealth is passed from families to private foundations for philanthropic purposes. Most academic researchers are not familiar with the aims or scope of private foundations, and developing this understanding is worth the effort. Support from foundations for exciting, creative ideas, particularly those developed "outside the box," is growing increasingly important.

The basic precepts for effective communication discussed in Chapter 3—organize, highlight, funnel, focus, and unify—hold true for all proposals, including those to foundations. Your application will be stronger if you adhere to the basic guidelines of knowing your audience, expressing a compelling vision, and demonstrating competence and ability to conduct the study. Avoid such common pitfalls as failing to establish the general significance, not linking your ideas to the specific work plan, overlooking critical information, or making a sloppy presentation. In fact, because foundation proposals are often much shorter than federal proposals, avoiding these pitfalls is even more crucial. It reminds us of the saying, "I didn't have time to write an article, so I wrote a book." That philosophy will not work with applications for foundation funding.

Although many general aspects of crafting a proposal to a foundation are similar to those for federal agencies, there are noteworthy differences. And there are no templates for designing effective foundation proposals—foundations' missions and criteria are extremely diverse. We urge you to contact the office at your institution that handles foundation relations, because this landscape is volatile and opportunistic. Foundations often rely on personal relations, which take time and foresight to develop. You may need to begin to put these in place well before you submit your first foundation proposal.

Foundation Basics

Private foundations are generally charitable entities established by individuals, families, or organizations, such as corporations with specific goals and missions. Foundations control a pool of money known as an endowment that generates income used to support the goals of the foundation. In the United States, foundations are defined by Internal Revenue Service tax code and are required by law to give away a certain portion of their funds (currently 5 percent) each year. They usually have very specific guidelines about the topics, organizations, and individuals they fund. At present, the Gordon and Betty Moore, Bill and Melinda Gates, David and Lucile Packard, Alfred P. Sloan, and Robert A. Welch Foundations are among the top funding sources specifically targeted for scientific, health, technology, and environmental research (see table).

The best rule to remember is that you cannot

Selected Foundations with Total Dollar Amounts and Numbers of Awards
in 2005

Foundation name	State	Dollar amount	Number of awards
Gordon and Betty Moore Foundation	CA	$92 million	48
Bill and Melinda Gates Foundation	WA	$92 million	18
David and Lucile Packard Foundation	CA	$36 million	25
Alfred P. Sloan Foundation	NY	$31 million	181
Robert A. Welch Foundation	TX	$24 million	141
Intel Foundation	OR	$10 million	57
M. J. Murdock Charitable Trust	WA	$6 million	49
Research Corporation	AZ	$5 million	·107

Source: The Foundation Center

generalize about foundations, but we are going to try
anyway:

- They like to fund innovation. They want to
  make a difference.

- They rarely award money to projects that could
  be funded by federal or state agencies.

- They are not required to offer detailed descriptions of their review process or follow the same
  procedures every year for awarding money.
  They have more latitude for rapid change than
  do federal agencies.

- They often have specific objectives for the projects they will support, yet these may not be
  stated fully in any written or online documents.

- Some have programs with fixed due dates, but others accept unsolicited letters of inquiry or applications at any time. Others will not accept unsolicited queries.

- Some will not work directly with individuals or sole-investigator projects, whereas others prefer to do so.

- Proposals are often short, sometimes in the form of a letter, so persuasive writing is crucial to your success.

- They often fund the person, so your reputation can play a decisive role in the success of your application. This also means that a poorly written proposal could result not only in denial of a specific request, but of all submissions from you (or your institution) in the future. Some foundations have long memories, so beware.

- Personal meetings with foundation personnel may be needed before a solicitation will be accepted, so prepare to spend time laying the groundwork for your application.

- Many institutions coordinate all requests to foundations. Discuss your ideas and targeted foundations with people in your institution before you make contact.

You can find out about specific foundations in many ways, such as conducting a Web search, visiting the Website run by the Foundation Center (www. foundationcenter.org), or keeping track of foundations mentioned in the acknowledgments of publications in your field. However, such material may not be detailed or current. The best way to get started with foundations is to talk to colleagues who have foundation support and to staff in your sponsored programs or foundations office. They can help you learn how to identify foundations that might welcome an inquiry from you or from others on your behalf.

How Do Foundations and Federal Funding Agencies Differ?

Like federal agencies, foundations usually fund nonprofit organizations (including colleges and universities) but also, less commonly, individuals or corporations. However, unlike federal agencies, foundations are more eclectic in their funding priorities. These priorities may change rapidly, for example, when there's a new foundation president or program staff member contributing new goals and objectives. Foundations also like to be flexible in order to meet what they perceive to be rising community, national, and international priorities that are not funded by other sources. The selection criteria and decision-making processes vary widely among foundations.

Private foundations are less likely to provide detailed descriptions of their proposal review procedures. Some foundations do not include indirect costs in their awards, while others allow a small percentage to be charged as "administrative costs." You will need to discuss this with your institution to determine its flexibility in policies regarding grants and adjustment of indirect costs. A few institutions discourage applications to foundations because of the possibility of lower indirect costs, but most welcome foundation awards, despite lower or no indirect cost recovery.

The types of initiatives or costs that will be covered by foundation awards vary widely. For example, some do not pay investigator salary. Some support travel costs, entertainment, living expenses, and per diem expenditures. Purchase of equipment, advertising for positions, consulting, and computational costs may not be allowed. For foundations, as a rule, you will need to clarify the acceptability of each type of cost you propose.

It is quite common for private foundations to require that their money be used to "leverage" other projects or costs. There are many ways to achieve this goal. You could show how foundation support will extend an ongoing project into a new area with a distinctly different and large payoff. You could also outline additional projects that you or your institution would develop as part of the proposed research, to be funded by another agency. Leveraging an investigator's salary is also very common; in this case, your institution covers the salary of the principal investigator as its contribution, and the foundation pays for other initiatives.

Foundations often allocate funds to institutions initiating new programs with the expectation that the institution will eventually support the programs on its own. Such conditions must be carefully considered, as increasingly foundations require evidence of institutional commitment to projects before making an award. Again, we emphasize the value of talking with administrators before submitting your proposal. These commitments can include matching funds, written assurances of plans to continue a program, and in-kind service. Your foundation officers should be able to assist you in determining the best way to meet these expectations.

Getting Started with Foundations

Spend some time with a senior colleague or a
person in your administration. Gather as much information as you can on your institution and its relationship with particular foundations. Often people do not
approach foundations until they have been funded by
more traditional funding agencies. Hence, it is often
the more established researchers who receive foundation support (except when it is not!). Foundation proposals always undergo some form of review—often
peer review—but the nature of that review is much less
well defined or transparent than those of federal agencies. Nevertheless, it is helpful to try to learn who the
audience will be for your foundation proposal (we gave
the same advice back in Chapter 1 for NSF-style proposals). In the case of a foundation proposal, it is unlikely—but it could happen—that your audience is
made up of entirely nonscientific staff. You should
know this before you start writing. Because of these
differences, especially where transparency and peer review are lacking, some private foundation funding may
not be considered as prestigious as federal agency
funding when tenure and promotion are in question.
We suggest that junior faculty talk to senior colleagues
or academic deans about the implications of receiving

foundation funding before spending too much time preparing a proposal. Also bear in mind that foundations with a known political agenda or particular desired outcome may carry less credibility among academics. Again, be sure to talk with colleagues to determine the reputation of a foundation and its suitability as a funding source, and how this might affect your professional position.

Exercises for Writing Proposals to Private Foundations

EXERCISE 7. (This follows from the exercises given in Chapters 4, 7, and 9.) The best way to begin is to consult with colleagues and administrators to identify foundations with interests that align with yours. In addition, read foundation Websites carefully, noting the topics and projects being funded, the size of awards, and geographic locations of funded activities. Read published outputs and annual reports from foundations where available. Eventually, you will want to contact a foundation officer before submitting a proposal, but as we have emphasized earlier, check with your institutional foundation office before doing so.

EXERCISE 8. Develop a number of ideas or perspectives on your work; the match between investigator and foundation may be highly dependent on the context you use to present the study. Your foundation staff may help you decide which parts of your proposal to emphasize or highlight when approaching different foundations. Sometimes this means looking at your project in a completely new way. We have colleagues who can take aspects of the same work and highlight different parts in order to appeal to NSF, NIH, and the American Heart Association, for example.

Most foundations fund scientific studies in several general categories. Consider how you could place your own work within each of these themes. Be aware that even these broad themes can change over time, so be sure to gather up-to-date information about each foundation to gauge its current areas of focus.

The four general research themes to consider are:

Region. Many foundations support the economic or social growth in a specific region (e.g., the northeastern states, the Colorado River watershed, western Pennsylvania, the Sierra Nevadas, the Sonoran Desert, etc).

Target group. This is a much broader category, and new additions arise according to societal pri-

orities. For example, in some years funding may
be very high for supporting K—12 educational
outreach programs, advancement of underrepresented minorities in science, junior faculty, or
mid-career advancement of women.

Topic. Some foundations have a general set of
guidelines and focuses, like cancer, diabetes,
sustainability, global change, poverty, disease,
etc.

Investigator or level. Many foundation programs
are designed to support junior investigators who
show outstanding potential and are selected in
large competitions. In these cases the investigator is usually nominated by an institutional representative, and quite often institutions can
nominate candidates only by invitation (often
limited to one or two nominees per institution).
Such awards are usually announced by the administration of your institution and may be
listed on a Web page in your sponsored project
office. Similarly, other awards are specifically intended for national and international leaders in
their disciplines. The merit criteria are usually
rigorous for these types of programs, and, as

such, they often are considered rather prestigious.

Finally, remember that the foundation landscape is constantly changing. Your options and chances could improve considerably from one year to the next, and we encourage you to reexamine foundation Websites periodically. We also emphasize the need to treat foundations and foundation officers with respect. These long-term, personal relationships can be extremely important and are quite different from those you develop with project officers at the federal agencies. Being careless or sloppy can have implications for your chances with private foundations long into the future.

Team Science for Tackling Complex Problems

Knowledge and access to information are expanding at an explosive rate—and this growth has increased the need for multidisciplinary teams of scientists to work on complex scientific problems. It used to be that a biologist working with a chemist was considered interdisciplinary. These days, scientists collaborate with experts in fields spanning the natural sciences, engineering, social sciences, and even the humanities.

Federal agencies and private foundations also are creating more programs to bring together multidisciplinary teams to address societal issues like climate change, cancer, and newly emerging, infectious diseases. Even within more traditional proposal review panels, investigators frequently are coauthoring proposals with colleagues outside their disciplines. For many scientists, multidisciplinary research—or so-called "team science"—is invigorating. Working with scholars from disparate fields who use different vocabularies and methodologies drives one to challenge one's own assumptions. Multidisciplinary collaboration is increasingly taught and then reinforced throughout undergraduate and graduate careers. We find tackling problems in this way highly rewarding and encourage our students and colleagues to do so when appropriate.

Writing multi-investigator and multidisciplinary research proposals, however, is more challenging and differs from writing more traditional single-investigator, single-discipline proposals. It takes special skill to construct compelling multidisciplinary research proposals, not to mention to be an effective member of a research team. Because of their breadth, successful proposals for team science depend heavily on effective language and tight organization. The more complicated the problems, the more links among aspects of the research, and the more people and institutions involved, the less space there is to elaborate on any individual idea. Every word becomes critical. Every project needs justification. And the gap between reviewer expertise and specifics of the work can be greater.

For a multidisciplinary proposal, a cohesive, compelling framework that unites the various pieces

"I'M ON THE VERGE OF A MAJOR BREAKTHROUGH, BUT I'M ALSO AT THAT POINT WHERE CHEMISTRY LEAVES OFF AND PHYSICS BEGINS, SO I'LL HAVE TO DROP THE WHOLE THING."

and underscores its greatest value is essential. To write persuasive multidisciplinary proposals, the fifth precept we introduced in Chapter 3 is particularly important: unify. Your reviewers will be asking themselves whether the sum of the work is greater than its parts. You will need to convince them that your questions warrant a team, that the study holds together, and that your team will be more successful and effective than individuals or smaller groups, evaluated separately. Special attention is needed to not only organize, highlight, funnel, and focus, but also to unify.

Why Multidisciplinary Research May Be Right For You

Working with a diverse team of collaborators enables one to ask complex and globally significant questions, include new techniques via collaborations, and synthesize across disciplines. The rewards are great, but the demand on time and resources will be considerable. There will be more people to manage, more complex finances, institutional rules, and additional research details to consider and track.

The National Academy of Sciences has con-

cluded that multidisciplinary thinking is "rapidly becoming an integral feature of research" (NAS 2005).
The NAS has identified four elements pushing this
forward:

- the inherent complexity of nature and society

- the desire to explore problems and questions
  that are not confined to a single discipline

- the need to solve societal problems

- the power of new technologies

Of course, these elements also are important in-
gredients of successful scientific proposals presented
throughout this book. Multidisciplinary proposals
share the majority of the elements that we have presented, so most of the chapters also pertain to collaborative proposals. Some agencies specifically ask for
collaborations across divisions and disciplines, and
sometimes investigators are asked to search for individuals to add a "human" dimension or a "science" aspect to a project. If you already are collaborating with
diverse groups, you are an ideal candidate for a multidisciplinary proposal.

There are also many types of multidisciplinary
proposals; even a single-author proposal can be, in
some sense, multidisciplinary. In this chapter we are

referring primarily to the proposals that make up collaborations from several investigators from different disciplines, often from different institutions or from different schools within the same institution. These proposals differ from more traditional proposals primarily in complexity and breadth, and they generally have larger budgets and may be funded for more years. We focus here on the issues posed by expanding breadth and administrative complexity.

Typical Multidisciplinary Proposal Requirements

There are several key areas that deserve particular attention when writing a multidisciplinary, multi-investigator, team science proposal.

- *Length and language.* Multidisciplinary proposals often have the same length constraints as other proposals yet typically need to cover more topics. This is possible only if the writing is succinct and well chosen. Carefully consider the need for each reference, and the amount of detail necessary for each method and analysis. Schematics and well-developed timelines help demonstrate how sections coordinate. Even a concise title that somehow conveys the breadth without complicating the work is valuable.

- *Unifying the voice.* In spite of their breadth, the best multidisciplinary proposals read as if they were written by one person. Parallel organization and consistent terminology help unify the voice and underscore the value of the collaborative approach and also help the reviewer keep it all straight. It takes extra work, but whatever you can do to make the proposal simple, elegant, focused, and easy to follow is worthwhile.

- *Institutional and program rules.* These proposals often have specific rules regarding the preparation of budgets, investigator credentials, and links between institutions. Coordinating rules across institutions often takes time and cooperation. There can be challenges even when collaborators come from the same institution if internal schools or programs have different indirect-cost rates or other administrative rules. Collaboration among large numbers of investigators often requires certain costs for grant management, building renovation, community outreach, and so on. Obviously, planning is essential. Understand as many of the rules as possible early in the process to avoid delays when finalizing or submitting the proposal. All large and most smaller institutions have staff in the grants and contracts office who can assist with these aspects of the proposal. The institution of the lead PI usually undertakes the final coordination of the submission.

- *Managing people.* As the number of people involved increases, the necessity for effective and transparent management becomes even greater. In many cases, investigators will come from different institutions, so the need to define lines of communication is particularly strong. As discussed in Chapter 2, establishing clear expectations for authorship, duties of the various investigators, and access to and use of the data will increase the success of the project. It will also reduce the chance for misunderstandings.

- *Ethical considerations.* Ensuring that everyone follows all rules (as discussed in Chapter 18) becomes more complicated as the program grows in personnel and intricacy of the connections. The PI ultimately is responsible for ensuring the integrity of the project and its management, but all project members should fully engage in such discussions. Transparent and open decision-making always increases the likelihood that the program is run efficiently and ethically.

- *Significance to merit the size and complexity.* Previous discussions of significance, objectives, and overarching goals have underscored the importance of these elements to the success of all research proposals. As the complexity of the problems or the size of the investigative team enlarges, so, too, must the significance. Connecting the actual work to the overarching goal is crucial. As reviewers, we have found this to be a common pitfall of team science. The question may be compelling and the team impressive, but what links the pieces is often underdeveloped. As a result, the vital element that makes the sum greater than its parts is missing, and the proposal is not successful.

- *Unifying the research.* Each part of a multidisciplinary project must be seen as essential to accomplishing the overall objective. The most effective proposals include a section that points out the links among sections that reinforce the overall research themes. These proposals frequently include plans for synthesizing the work explicitly in the program plan. Some funding agencies require a specific plan for coordinating projects.

- *Understand the reviewer process.* The National Academy of Sciences notes that certain funding organizations have been reluctant to fund multidisciplinary research because "it requires risk-taking and administrative complexities" that typical research does not require. For example, a traditional panel of reviewers will probably not be sufficient for reviewing multidisciplinary proposals, so even the task of assembling a panel becomes more complex. The panels reviewing multidisciplinary proposals must be diverse because they have to cover broad areas of research. It follows that many or even most reviewers will not be familiar with every field of expertise in the proposal. As we noted for foundation proposals, some of the reviewers also may be nonexperts. For these reasons we urge you to fully understand the reviewing community before writing your proposal, and to craft different sections of the proposal to be accessible to the different types of reviewers (e.g., methods sufficient for experts to evaluate, significance accessible to all reviewers). A strategic discussion with the panel director can be very helpful in learning the review process.

Exercises for Getting Started

The exercises presented in the other chapters are all appropriate for multidisciplinary proposals. The following drills are meant to assist you with getting collaborations under way and unifying research for the project.

EXERCISE 9. (This follows from the exercises given in Chapters 4, 7, 9, and 16.) Multidisciplinary research teams often consist of both experienced and new faculty. This offers great opportunity for mentoring while the research is being conducted. One option, especially for newer investigators, is to approach senior colleagues to ask whether they know of any such opportunities. It is much easier to start out as a co-PI or collaborator on a large project of an experienced investigator than it is to organize a project yourself. You can also talk to your academic administrators to identify in-

ternal programs that can support the development of
larger projects and collaborations. It is not uncommon
to apply for internal funds for pilot research before
bringing together a coalition of scientists to write a program project. Similarly, you may wish to develop a
symposium at a research conference or at your institution to bring together potential colleagues to think
broadly about a research topic. These activities often
lead to bigger collaborations, and they have merit on
their own for establishing exciting scientific connections.

Together with the investment from the funding
agency, investigator institutions are routinely expected
to make contributions for large multidisciplinary proposals. Determine at the start if the funding agency has
requirements for matching funds, cost sharing, graduate stipends, and other types of in-kind assistance, and
explore this with your home institution as well.

EXERCISE 10. Early in the process, develop a section that unifies the disparate aspects of the work. Such
unifying themes need to be reinforced in all sections of
the proposal. Other suggestions: construct a diagram
to make the links very clear; share the overarching significance with colleagues in a variety of fields to determine if the content is compelling even to nonexperts;
ask colleagues if the title provides insight to the larger

goals; make sure that every section is explicitly tied to the unifying themes in some fashion; use headers to unify sections; and include a precise plan to synthesize results throughout the project rather than waiting until the research has been completed.

Ethics and Research

The conduct of science assumes that the people involved are going to be honest and trustworthy (Macrina
2005). Therefore, you accept full responsibility for ensuring the integrity of your work when you conduct scientific research. If you accept financial resources to
fund your research, you also are fully accountable to
your institution, the scientific community, and the
funding agency. Anyone who enjoys the privilege of
leading a laboratory has a further responsibility to
make certain that their students and colleagues also
understand the expectations and standards required
(Shrader-Frechette 1994).

In this chapter we point out a few topics related
to ethics that should be understood by all graduate students, faculty, and investigators involved in scientific
research. We urge you take a class or participate in a
seminar or reading group on this subject at some point

in your career. Most research institutions offer or even require such courses of graduate students, and certain funding agencies require that all of their recipients complete one.

Issues to Consider

Many ethical challenges can arise when one is conducting scientific research with public or private funds. Some of these, such as proper citation techniques, ownership of ideas, and budgetary oversight, have been discussed in other chapters. In this chapter we identify five general areas in which ethical standards must be maintained. Reflecting on these vital aspects of conducting research is a healthy exercise. Our own students and colleagues enjoy talking about these issues, and our programs have benefited from such conversations at all phases of proposal development and implementation.

Give appropriate credit. As you write your proposal, and as you present your ideas to others informally or in seminars, recognize the contributions of others. It is always appropriate to acknowledge that certain ideas in your proposal derive from the work of others. As we discussed earlier, giving credit can reduce the chance for more complicated problems, such as authorship disputes and the need for scientific oversight to maintain the integrity of the data. Early, frequent, and open communication significantly reduces the possibility of disputes among investigators.

Respect people, animals, plants, and the environment affected by your research. Federal, state, and institutional regulations to ensure this have become common. All academic institutions have committees that routinely review research involving human subjects and animals. Many funding agencies require specific information on animal care and maintenance to be included in the grant application (see Chapter 3). However, even if your research does not fall under these regulations, try to minimize unnecessary impacts of your activities. A discussion of this with colleagues could be useful, especially for beginning researchers and graduate students.

Remain objective. Although no one starts out intending to falsify data, scientific misconduct occurs and is always devastating to the scientific community. Remaining objective, not preguessing your results, and sharing with others your decisions about data analysis and interpretation are the best ways to avoid such problems. For example, a researcher may become so convinced of the outcome of an experiment that he or she ignores or dismisses contradictory results. Or an investigator may choose to conduct experiments that favor particular outcomes. By discussing these decisions openly you may reduce the chance of biasing your approach or results. Scientists have many ways to ensure the validity of research (e.g., peer review, publishing results and data), but it is far better to avoid such problems than to have them discovered in the review process. Sharing your ideas and research designs, encouraging and soliciting criticism, and being self-critical will ensure that you remain objective throughout your research career.

Spend money appropriately. Accepting financial support for research constitutes an agreement that you will spend the award money on the approved research and that you will follow all guidelines for reporting your use of funds. Different agencies have specific regulations concerning the shifting of money from one budget category to another. Deviations from the original, approved budget often require explanation, justification, and permission from the agency. Expenditures on nonresearch items, or wasteful spending, are never permissible. Financial misconduct is not tolerated, regardless of intent or even understanding (it is not a defense to say that you did not understand), so maintain

---

Integrity in Research

The individual scientist agrees to the following when he or she undertakes scientific inquiry:

- intellectual honesty in proposing, performing, and reporting research

- accuracy in representing the contributions of individuals to developing and writing research proposals and to subsequent reports and publications

- collegiality in scientific interactions, including oral and written communications and use of resources

- protection of human subjects, humane care of animals, and responsible treatment of the environment

- respect for the individual and collective responsibilities of investigators and their research groups

Source: Modified from "Integrity in Scientific Research," National Research Council (2002), p. 5

---

close relationships with your granting officer and ask questions whenever you are unsure of regulations.

Be attentive to possible conflicts of interest. Identify any COI—or any situation that might even appear to be a conflict of interest—to the appropriate people at your institution before you submit your proposal. A COI typically occurs when a person's private interests interfere or conflict with the obligations or responsibilities that person has in a public capacity, usually related to a job. If you have personal financial interests (stocks, investments, etc.) that might benefit from the outcome of research you propose to do, you should definitely talk to the appropriate people at your institution about a possible conflict of interest. When you work with a close relative, a partner, or someone with whom you have a deep personal or romantic relationship you should consult with your institution about proper reporting pathways. Most institutions have guidelines that can be very useful and will assist you with avoiding problems associated with COIs.

Use the Guidance and Advice of Others

Our strongest recommendation is that whenever you have ethical questions—whether you are a first-year graduate student or an experienced researcher—the best approach is to talk to colleagues and mentors in your field, and to officials at your institution who handle these types of issues. Discuss your potential difficulties with those you respect and you will almost certainly avoid seeing them become dilemmas.

References

Bacchetti, R., and T. Ehrlich (eds.). 2007. Reconnecting Education and Foundations. Wiley, San Francisco.

Day, R. A., and B. Gastel. 2006. How to Write and Publish a Scientific Paper (6th ed.). Greenwood, Westport, Conn.

Glass, S. A. (ed.). 2000. Approaching Foundations. Jossey-Bass, San Francisco.

Hacker, D. 2003. A Writer's Reference (5th ed.). Bedford/St. Martin's, Boston.

Hartsook, R. F. 2002. Nobody Wants to Give Money Away! ASR Philanthropic, Wichita.

Locke, L. F., W. W. Spirduso, and S. J. Silverman. 2000. Proposals That Work (4th ed.) Sage, Thousand Oaks, Calif.

Longman, A. W. 1998. Author's Guide. Addison Wesley Longman, New York.

Lunsford, A. A. 2005. The Everyday Writer (3rd ed.) Bedford/St. Martin's, Boston.

Macrina, F. 2005. Scientific Integrity (3rd ed.). American Society for Microbiology, Washington, D.C.

Mathews, J. R., J. M. Bowen, and R. W. Matthews. 1996. Successful Scientific Writing. Cambridge University Press, Cambridge.

Mathews, J. R., J. M. Bowen, and R. W. Matthews. 2000. Successful Scientific Writing (2nd ed.). Cambridge University Press, Cambridge.

McCabe, L. L., and E. R. B. McCabe. 2000. How to Succeed in Academics. Academic, San Diego.

National Academy of Sciences. 2005. Facilitating Interdisciplinary Research. National Academy, Washington, D.C.

National Research Council. 2002. Integrity in Scientific Research. National Academy, Washington, D.C.

Ogden, T. E., and I. A. Goldberg. 2002. Research Proposals (3rd ed.). Academic, San Diego.

The Online Ethics Center for Engineering and Science at Case Western Reserve University. http://online ethics.org/; viewed in 2006 and 2007.

Proctor, T. 2005. Creative Problem Solving for Managers (2nd ed.). Routledge, New York.

Reif-Lehrer, Liane. 2005. Grant Application Writer's Handbook (4th ed.). Jones and Bartlett, Sudbury, Mass.

Runco, M. A. 1994. Problem Finding, Problem Solving, and Creativity. Ablex, Norwood, N.J.

Shrader-Frechette, K. 1994. Ethics of Scientific Research. Rowman and Littlefield, Boston.

Wang, O. O. 2005. Guide to Effective Grant Writing. Springer, New York.

Ward, D. 2006. Writing Grant Proposals That Win (3rd ed.). Jones and Bartlett, Sudbury, Mass.

Index

Susan Milord

Andrew J. Friedland is an environmental scientist and the chair of the Environmental Studies Program at Dartmouth College. He has conducted long-term studies on lead cycling and mobility in high-elevation forests of New England. He has received funding from NSF, USDA Forest Service, EPA, and private foundations. Friedland is the Richard and Jane Pearl Professor in Environmental Studies.

Kawakahi K. Amina

Carol L. Folt is an environmental biologist and the dean of the faculty of arts and sciences at Dartmouth College. She is associate director of Dartmouth's multidisciplinary Superfund Toxic Metals Basic Research Program and is cofounder of the Center for Environmental Health Sciences at Dartmouth. She has received funding from NSF, NIEHS, EPA, USDA, DOE, and private foundations. Folt holds the Dartmouth Professor Chair in Biological Sciences.